W9-BZY-223

bayshore
summer

Houghton Mifflin Harcourt · *Boston* · *New York* · 2010

bayshore summer

FINDING EDEN IN A MOST UNLIKELY PLACE

PETE DUNNE

Photographs by Linda Dunne

Copyright © 2010 by Pete Dunne
Photographs copyright © 2010 by Linda Dunne

All rights reserved

For information about permission to reproduce selections from this book,
write to Permissions, Houghton Mifflin Harcourt Publishing Company,
215 Park Avenue South, New York, New York 10003.

www.hmhbooks.com

Library of Congress Cataloging-in-Publication Data

Dunne, Pete, date.
Bayshore summer / Pete Dunne; photographs by
Linda Dunne.
p. cm.
Includes bibliographical references.
ISBN 978-0-547-19563-6
1. Natural history—New Jersey—Cumberland County. 2. Summer—
New Jersey—Cumberland County. 3. Outdoor life—New Jersey—
Cumberland County. I. Title.
QH105.N5D86 2010
508.749'4—dc22   2009027928

Book design by Anne Chalmers
Printed in the United States of America
DOC 10 9 8 7 6 5 4 3 2

To Rich and Pat and Karl
and Gordon and Tom

Acknowledgments

As is common with writing projects, a number of fine individuals contributed their time and talents to this one. They have my unqualified gratitude, and I am delighted to introduce them to you.

Robert H. Francois of the Cumberland County Historical Society; Alan Carman, curator of the Cumberland County Prehistorical Museum; and Irene Fergeson of the Mauricetown Historical Society, whose extensive knowledge and guidance figure in this book.

Phil Correll of the Coastal Heritage Trail and Jane Galetto of Citizens United to Protect the Maurice River and Its Tributaries; Dale Sweitzer of The Nature Conservancy and Kim Wood of Cumberland County Tourism for counsel and guidance regarding subjects in their fields of expertise.

My colleagues Don Freiday and Deb Shaw for reviewing the manuscript and offering many helpful suggestions.

My agent, Russell Galen, whose specialty is making it easy for writers to focus on what they really want to do (which is write), and my longtime editor at Houghton Mifflin Harcourt, Lisa White, who continues to make all of our projects together a pleasure.

And, of course, finally, and needless to say, Linda, who figures in everything and whose photos make the words I write supportive at best.

Contents

PROLOGUE:
Memorial Day Weekend; Bunting

The sun, changing now from Jersey tomato red to California poppy gold, was smack in the windshield as our car crested the bridge and lined out along the causeway, heading for the distant wall of trees.

What bridge?

The bridge over the Maurice River. The only bridge spanning this federally designated wild and scenic waterway south of Millville.

Millville?

The old South Jersey industrial town built on, and from, the sand that constitutes New Jersey's coastal plain and makes up about half the state.

If you were a local you'd know these things.

"Osprey," Linda announced, nodding in the direction of Delaware Bay. Sure enough, high over the marshes, one of our local "fish hawks" was pinned to the sky. Even through closed windows, we could hear the strident piping of its territorial call—in the language of ospreys, a notice to neighbors and would-be neighbors that this stretch of marsh was his and theirs was somewhere else.

Thirty-two years ago, when I first crossed this bridge, heading west, not east, there were no ospreys nesting here. On New Jersey's Delaware Bayshore, as elsewhere, the population of these golden-eyed raptors had been ravaged by DDT. Now, after decades of recovery, more than thirty-five established pairs nest along this forty-two-mile waterway, with more prospecting birds appearing annually. Hence the need for resident birds to post notice of occupancy, or, to paraphrase Robert Frost, good displays make good neighbors.

"Uh-oh," Linda observed.

"What did you forget," I said more than asked.

"My bug shirt," she said. "Left it on the kitchen table."

If you are impressed by my apparent prescience, you needn't be. Linda and I have been married for more than twenty years. You learn a great deal about the workings of another person's mind with that combination of longevity and intimacy.

What's more, at that very moment we were setting off to begin another adventure together: a new book. You are guaranteed to forget something important when you start a project this big. And on New Jersey's Delaware Bayshore, in late May, on a windless day, a bug shirt is not just important—it can mean the difference between life and death.

"We'll go back for it," I said.

"I don't want to lose any tide," Linda objected, and, no, this isn't a typo. Here on the bayshore, lives are governed by the daily rise and fall of the tide rather than the erosive and unrelenting race with time that seems to govern the rest of the world. The affirming allure of this synchronous rhythm goes a long way toward explaining why people who are born here have a difficult time leaving and why transplants, like Linda and me, came and stayed.

"Don't be silly. We're only two minutes from home." On most of our projects—including *Prairie Spring*—Linda and I have been far from home. Any item left on the kitchen table, however critical, was destined to remain there until we returned, weeks or months later.

This project is different. This book is about home, the place we have chosen to live our adult lives.

In less time than it takes a territorial osprey to conclude his display, we retrieved the missing bug shirt, recrossed the bridge, and nosed our Subaru Outback into the local Wawa to get gas.

What's a Wawa?

It's Lenape for 7-Eleven, a convenience store chain known throughout the Philadelphia region for its coffee, hoagies, and inexpensive gas. *Wawa*, in the language of the Lenape people, means "goose," and the Canada goose is the emblematic bird of the Wawa chain.

The Lenape (pronounced *Len-op-ay*) were the native peoples whose encampments dotted the region when Europeans first reached these shores. In fact, the Maurice (pronounced *moor-us*) River is named for a Dutch ship, the *Prince Maurice,* which legend has it was boarded and burned by a Lenape raiding party in 1693.

Yep, and for much the same reason ospreys display over their territories, a cautious reserve regarding newcomers has a long-standing tradition on the Delaware Bayshore.

See, See, see, see, see

At the Wawa, all islands and pumps were operating, but on this busy Saturday of Memorial Day weekend, only two spots were free. We slipped into one of these, just ahead of a black SUV with Pennsylvania plates towing a twenty-foot Boston Whaler.

Linda complains that every other car on the road in New Jersey is black, but she's wrong. My polling shows one in five. Four at the most.

"This place is humming," I observed.

"They're heEEEer," Linda chanted, mimicking the line made famous in the movie *Poltergeist.*

The evening before, on my way home from New Jersey Audubon's Cape May Bird Observatory, the place I hang my director's hat, the backup of shorebound travelers had been four miles long. Fortunately, I was heading the other way.

Three hours later, when Linda finished her nursing shift in the emergency room at Cape Regional Medical Center, the line had stretched six miles—all the way to, and beyond, the Cumberland County line.

Weekend traffic jams on Route 347 and parallel Route

47, the roads most favored by Philadelphia residents en route to Cape May beaches, are the summer norm. But on holiday weekends, the volume of traffic about doubles.

The funny thing—the sad thing—is that all those frustrated drivers are not only wasting quality time but missing the boat. Every Friday and Sunday between Memorial Day and Labor Day, these tens of thousands of recreation-seeking travelers drive right past one of the planet's best-kept secrets. A land of unencumbered space and natural beauty. New Jersey's forgotten shore.

"Fill it; regular," I said to Amanda, a pert, young fuel pump attendant and resident of nearby Port Norris—a backwater coastal community known a century ago as the "Oyster Capital of North America." Back when Port Norris boasted of having the highest number of millionaires per square foot in the United States, Amanda's grandfather was the editor of the town newspaper. It published its final edition soon after the oyster industry collapsed, and the town, along with the region, went into economic suspension.

"It's jammin' here," I noted conversationally.

"Yep," Amanda said, making a tightlipped expression that was half-grimace, half-grin. "You want to see crowded, you stick around till nine."

"Why nine?"

"It's when I go on break," she said, jamming the nozzle in the Subaru's tank. "And we shut down some pumps."

We didn't have time to wait. And crowds were precisely what we had a mind to avoid.

"I'm going to get coffee," I told Linda, whose face adopted

an expression that was somewhat like Amanda's but imbued with less grin. The expression a photographer en route to a thematically important and tide-sensitive subject might make when her partner says he's going for coffee.

"Relax. I'll be back before the tank is full," I promised, and it was a promise I almost kept.

Halfway across the parking lot, my passage was halted by the ethereal song of a blackpoll warbler emanating from the bordering forest. A Neotropical migrant and northern breeder, the blackpoll ranks among the last of the spring migrants to pass through our region.

The song is high-pitched, the notes breathy. Phonetically rendered, it seems as if the bird is urging listeners to *See, See, see, see, see* . . . with the last, frail notes evaporating into high, thin air. Over the many years I've been attuned to such things, the song of the blackpoll has come to represent to me the sound of Taps being played for spring migration.

"Hear the blackpoll?" I shouted toward the car.

Linda pointed to her watch.

Did I mention that our Subaru Outback is black? Do I have to tell you that I picked it out?

That's right. I'm a New Jersey native, born and bred. Linda? She's originally from Southern California (where every other car is white, and time is measured in how many minutes it takes to drive someplace else).

SUMMER 101

Inside the Wawa it was bedlam. The place was jammed with shorebound visitors dressed for summer despite the early morning chill as well as local roofers, painters, plumbers, and laborers who had been transformed into fishermen, backyard barbecue chefs, and horseshoe-pitching champions by the magic inherent in a three-day weekend.

Of course Memorial Day is not summer—not strictly speaking or in fact. Summer, like all seasons of the earth year, is a period that relates to, and is defined by, the annual dance of our planet around its star. It takes 365 days for the earth to complete its circumnavigation of the sun. In this span of days, our planet, and its inhabitants, will pass through four seasons—each of which is about 91 and ¼ days, or three months long.

Winter, spring, summer, autumn.

You know all this.

What you may not know is why this annual trip around the sun results in seasonal shifts between warm and cold. While it has a great deal to do with the heating rays of the sun, it has little to do with the sun's energy output or the spatial relationship between our star and the third planet in its system of eight orbiting bodies.

The planet we live on. Earth.

In fact, the seasonality we experience has everything to do with the earth, more specifically the angle of the earth and how it presents itself to its energy source, the sun. But while this geometric interplay between the earth and its star is very important to life on Earth, and arguably to books whose focus is the seasons, it is also fairly boring. What's more, as regards twenty-first-century Americans and our sense of summer, it is also pretty irrelevant.

What we call "summer" is more market-contrived than astronomically correct.

So let's supersimplify this. Let's limit our discussion of seasonality, beginning with the key understandings that our spinning planet rotates on an axis and that the axis doesn't run straight up and down. It is tilted. What this means is that, as our planet orbits the sun, the upper and lower halves will lean, alternately, toward and then away from the sun.

Half the year leaning toward, half the year leaning away.

When you, and that portion of the planet you are standing on, are inclined toward the sun, you're in the warming season, whose core is summer. The people in the Wawa respond to this by buying cold soft drinks, ice cream, and Italian hoagies.

Hoagie, incidentally, is South Jersey for "sub sandwich."

No, the Lenape didn't name it.

When you and your hemisphere are leaning away from the sun, that's the cold season, or winter. People going into the Wawa are buying coffee, hot chili, and Philly cheese steaks.

As our planet orbits the sun, the actual tilt or angle of the earth on its axis doesn't change, but its angle relative to the sun does. From plus twenty-three degrees to minus twenty-three degrees in six months' time. The times when the planet's two poles are leaning, respectively, closest to (plus twenty-three degrees) and farthest from (minus twenty-three degrees) the sun are the solstices—the first day of either the summer or the winter (again, depending upon which hemisphere you live in).

In the Northern Hemisphere, the actual first day of summer falls on June 20 or 21. On this day the sun is twenty-three degrees above the horizon at the North Pole, smack overhead at the Tropic of Cancer, and at its highest point in the sky over Cumberland County, New Jersey (and every place else north of the Tropic of Cancer).

So why has Memorial Day, celebrated in late May, come to be the traditional first day of summer when the actual first day of summer is still nearly a month away? A combination of good weather, good marketing, and creative legislating.

Across much of northern North America (including Cumberland County), late May is temperate enough to support outdoor activities. The Northern Hemisphere has, after all, been in a warming trend since one second after the winter solstice—an event that transpired way back on December 21 or 22. After five months of ever-increasing solar insolation, the earth and surrounding air have gotten pretty cozy.

Along the Jersey Shore, water temperatures are still in the fifties or sixties (too cold for swimming), but places whose economy depends upon tourism have a vested interest in pushing the boundaries of the season, and, heck, who doesn't like a holiday?

That's where the creative legislating comes in. In 1971, Congress passed the National Holiday Act, changing the celebration of Memorial Day from May 30 to the last Monday in May and creating a federal holiday—a change that ensured Americans an annual three-day, getaway weekend every year (instead of two out of every seven).

Originally known as "Decoration Day," a day of remembrance for the fallen of the Civil War, the holiday was first celebrated in 1868 at Arlington National Cemetery, where flowers were placed on the graves of both Union and Confederate soldiers.

Incidentally, this bipartisan recognition resonates well in Cumberland County, New Jersey. While officially part of the Union during the Civil War, many residents of agricultural South Jersey had more cultural affinity (and in some cases, loyalty) to the Southern cause. A few of the young men from South Jersey even fought for the Confederacy (in fact, the flag from the *Merrimac*, salvaged by a Cape May native and member of the crew when the Confederate ironclad was scuttled, resides in the Cape May County Historical Museum).

Across most of North America, Memorial Day, as it became popularly known after World War II, is celebrated by the planting of flags upon the graves of American war veterans, speeches, parades, and the sale of poppies by the VFW to aid disabled veterans.

It is also the weekend of the Indy 500 car race, a near epidemic of yard sales, backyard barbecues, the kickoff for the national Click It or Ticket campaign, and the semiofficial start of the Summer Vacation Season.

A season that lasts until another legislatively contrived and noncelestial-based holiday, known as Labor Day.

In twenty-first-century America, the actual first day of summer, falling on June 20 or 21, is mostly unheralded, uncelebrated, and overlooked.

WHAT, WHAT? WHERE, WHERE?
HERE, HERE. SEE IT. SEE IT.

Because of the lines at the checkout counter, my coffee run took longer than planned. Exiting the Wawa with two cups of coffee in hand, I noted, first, that the gas nozzle was no longer stuck in our Subaru's tank (oops!) and, second, that even at this early hour, traffic was building.

I smiled. Unlike other Wawa patrons, we weren't going to be affected by the traffic.

Over the sputtering grunts of a dozen Harley-Davidson motorcycles and the wail of a preschooler who hadn't quite made it to the Wawa restrooms in time, my ears picked out the glittering notes of an indigo bunting's song hailing from the brushy lot across the causeway. Among the planet's most stunning birds, the bright blue males put a summer sky to shame, and the high, clear notes of this summer resident's song sparkled like audible sunlight.

Phonetically rendered, it sounds as if the bird is exhorting: *What, what? Where, where? Here, here. See it. See it.*

Not for the first time, and not just because the bird called the question, I wondered how many of these shorebound commuters had any inkling that they were a single right-hand turn away from a world of discovery and wonder.

I got in the car. Muttered something about longer-than-usual lines. Waved to Amanda. Pulled away from the pumps to make room for a guy towing a pop-up camper, looking at his watch, and wearing a world-class frown. Turned east onto old Route 47 untroubled by the thought of the traffic jam that lay ahead.

We'd be getting off well before any backup, taking the indigo bunting's advice. *What* Linda and I planned to do was witness one of the planet's greatest natural spectacles. *Where* is one of the secrets that will be disclosed in this book because *here,* on New Jersey's forgotten coast, are wonders enough to fill a season and spill into the next.

The day was beautiful. The tide was perfect. And we were going to *see it, see it.*

CHAPTER 1

Sex and Gluttony on Delaware Bay

Before my hand found third gear, New Jersey's famed pine barrens had closed in on both sides of the road, buffering us from the Wawa and its mayhem. Much of eastern Cumberland County falls within the boundaries of the Pinelands National Reserve—a 1.1-million-acre tract of mostly forested land. Not a park or refuge where human endeavors are strictly proscribed, the reserve is a jurisdictional hybrid according protection to areas of high environmental or cultural value and permitting compatible development in others. Created by Congress in 1978, operating under a comprehensive management plan, and governed by the representatively diverse Pinelands Commission, both the reserve and the plan are variously acclaimed and decried.

People who want to see the environmental and cultural heritage of the region preserved generally applaud it. Large property owners and developers whose ambitions might undermine environmentally sensitive areas commonly fault it.

Like it or not, what is certain is that this creative initiative resulted in the protection of 22 percent of the most crowded state in the Union and preserved the largest body of open space between Boston and Richmond. Those shore-bound travelers taking old Route 47, as Linda and I had done, are tracing the southern boundary of the preserve.

For some reason, the Delaware Bayshore was not placed beneath the umbrella of the reserve's protection. Its heritage and biological riches have, thus far, been preserved largely as a result of good fortune.

We passed several turnoffs, marked by weathered signs, directing travelers to communities with names that skirt all but local recognition—names such as Dorchester (pronounced, in the syllable-compressing dialect of the bayshore, *DOR-ster*), Leesburg, and Heislerville. While few travelers ever do turn off to explore these obscure towns, those who do wander into a New Jersey that is alien to nine out of ten residents of the state.

A land of black ducks and blue mud. Tight-knit communities composed of cedar-roofed and dowel-framed houses that hark back to a time when two-masted schooners and two-story homes were built by the same men.

Pickup trucks, not cars, occupy most of the driveways.

Cats keep watch from behind age-rippled glass, and dogs, slumbering on porches, stir only at the sound of unfamiliar feet. And when you happen, as is certain you will, upon one of the old church cemeteries that command the high ground, you'll find the gate unlocked and age-blackened stones bearing the same names as the mailboxes you just passed.

Until recently, saying you lived on the bayshore was as good as saying you were born on the bayshore. This means you have roots that go back very far.

Not every road off Route 47 delivers as promised. Travelers bold enough to follow now rusting highway signs to Moores Beach and Thompsons Beach are headed for consternation. In two decades, these bayside communities went from shore towns to ghost towns to no towns as the roads leading out to them surrendered to marsh and the houses fell (literally) under the dominion of the tide.

Today, if you own a four-wheel-drive vehicle that you don't mind marinating in salt water, you can still drive to Moores Beach and the rubble-strewn strip of sand that borders the bay and once supported a town. But the road leading out to Thompsons Beach is impassable and gated. Beneath its blanket of salt grass there is macadam. At low tide it is navigable to foot traffic (or, more accurately, knee-high-boot traffic). But under the very best of conditions, the road that once led to this fishing community is treacherous. A person (or persons) would have to be very foolish, or very motivated, to attempt it.

BLUE MUD

"Eahhh!" Linda cried, the sound of her protest barely audible over the cries of laughing gulls and the belly-laugh grunts of clapper rails—both of which are common summer residents.

"What?" I asked, turning, looking back at my wife, who was swaying, ominously, at a better than twenty-three-degree list, in the middle of a puddle that was the size and depth of a kiddie pool, though a good deal muddier.

Why ominously? Because the pack on her back was crammed with about twenty thousand dollars' worth of camera equipment and, as the Nikon owner's manual cautions, digital cameras and salt water don't mix.

Linda's no stranger to portaging stuff on her back. A onetime National Outdoor Leadership School instructor, she once climbed Denali wearing a hundred-pound pack—a pack that weighed as much as she did. Concluding, therefore, that ballast was not the source of my wife's consternation, I explored the possibility of an equipment malfunction.

"Boots leak?" I asked, when her gyrations had stabilized enough to support conversation.

"No!" she shouted. "They're too large. The muck keeps sucking them off my fee——*Ehh; ahhhh.*"

I tensed as Linda survived another arm-swinging battle with gravity. Search the world over, you'll find hardly anything more slick, slimy, and boot-suckingly treacherous than good ol' Delaware Bay blue mud. The fine particulate matter, ferried and deposited by the waters of the Delaware River, has the color and consistency of graphite and the sulfurous smell of hell. Dark spatterings of the stuff were already marring Linda's pretty face. But since she wasn't going to be on the receiving side of the cameras she was carrying, it hardly mattered.

"We're almost there," I encouraged. "The road's better ahead." This was true as far as it went. What I didn't say, which was also true, was "and then it gets worse again."

Ten minutes later, our boots, still numbering four, were planted on packed white sand. In front of us was a sunsplashed Delaware Bay. Around us the rubble that used to be Thompsons Beach.

"There are no birds," Linda couldn't help noticing.

"Well, there were," I asserted.

"When?" she wanted to know.

"About twenty years ago," I said, smiling quickly to let her know it was a joke.

"This way," I said, turning east. "Another quarter mile. I scouted it yesterday. An hour earlier. The tide should be about the same, and there absolutely were birds."

"Including knots?" she asked, pinning a name to the poster bird of Delaware Bay's famed spring shorebird concentrations.

"Including knots," I said. "Some. Ready to go?"

She was, and we did. Walked east along the narrow strip of sand and through the remains of the town. But I couldn't help thinking of the way it had been twenty years ago—

both with the town and with the birds. And how the diminishment of what was one of the planet's greatest natural spectacles was anything but a joke.

CRABS AND SHOREBIRDS FOR LUNCH

My introduction to the now-famous spectacle of spawning crabs and migrating shorebirds came in May 1977. I was responding to an invitation by Jim and Joan Seibert, residents of the bayside hamlet of Del Haven—a cluster of houses clinging to the land about halfway up the Cape May peninsula.

They said that the beach in front of their house was awash in birds.

"Fantastic," they assessed. "Unbelievable," they promised. "Come to lunch."

I did. And while, as the newly minted naturalist for the incipient Cape May Bird Observatory, I would have come just to see the birds, the promise of a free lunch was irresistible.

Naturalists the world over are opportunistic feeders. Since my position with the New Jersey Audubon Society was earning me a whopping $250 a month, such flexible feeding habits had clear survival advantages. In this regard, naturalists are much like the shorebirds that concentrate here in spring. They, too, are in it for the eats.

Lunch was enjoyable, the birds as advertised. Fantastic, unbelievable. A pulsing, vibrating ribbon of shorebirds stretching as far as the eye could see—tiny gray-backed semipalmated sandpipers, sanderlings in their rarely seen brick red breeding plumage, harlequin-patterned ruddy turnstones, and, best of all, red knots! Big, burly, silver-backed shorebirds and high-Arctic breeders. There were more knots on that single stretch of beach than I'd ever dreamed of. More knots, as it turned out, than were estimated to be in all of North America.

But just as impressive, perhaps more, was the concentration of breeding horseshoe crabs, whose tiny gray-green eggs were the foundation and objective of the feeding birds—the tiny loaves that fed the multitudes.

Hubcap-sized, bronze-colored, and helmet-shaped, the 340-million-year-old sea creatures absolutely carpeted the beach. There were hundreds of larger female crabs, half buried in sand, and thousands of attending males.

To the crabs, it was all about sex. To the birds, it was all about gluttony.

Many years later, I would read an account of the spawning crabs written by Alexander Wilson, the father of American ornithology. Writing in the early 1800s of the stretch of beach east of the mouth of the Maurice River—the stretch of beach Linda and I were walking now—Wilson noted that a person could walk ten miles upon the carapaces of horseshoe crabs and never touch the sand.

It was this way, too, in 1977 and for about a decade after, and my experience here is firsthand.

Starting in 1981, I was part of a three-person team flying aerial surveys to gauge the magnitude of the shorebird concentrations along New Jersey's bayshore. We counted a peak one-day total of 350,000 birds during those inaugural survey flights, including 67,000 red knots. The following year, adding the cross-bay beaches of the state of Delaware to the survey route, our totals reached 420,000 birds, including 95,000 red knots.

Given a turnover of fourteen days, biologists projected that a total of a million to perhaps a million and a half shorebirds were foraging on bayshore beaches between May 10 and June 10, making the spectacle on Delaware Bay one of the greatest concentrations of shorebirds on the planet. From there, fully fueled on the fat of reconstituted crab eggs, the birds would fly nonstop to Arctic breeding grounds and arrive in fine shape to go about the serious business of replicating the species.

The crabs were estimated to number 10 to 20 million, by far the greatest aggregation of this living fossil known to science. On every high tide, when the crabs would emerge, the shells in the surf sounded like crockery rattling in some great sink. As the tide receded, it looked as though the beaches were paved in animate cobblestones.

That was the celebrated phenomenon. An annual massed gathering of living things on the narrow beaches of Delaware Bay. Nobody knows how long it had been going on. But everybody knows how long it lasted after it was rediscovered. About fifteen years.

KNOT NOW

Past the pilings that once supported the homes and docks of Thompsons Beach, and the rubble that is all that is left of them, there is a quarter mile of open beach. The tide was still rising. Here and there along the old high-tide line were shallow craters that marked the location of newly deposited horseshoe crab eggs. Every ten yards or so, Linda and I encountered an overturned crab—mostly male. We righted these with our boots, to the annoyance of the onlooking host of herring gulls, who enjoy scavenging rights to all dead and dying crabs.

The winds, light and northerly, were projected to go slack, then southerly by afternoon. Good news for crabs and birds primed to migrate. Not so good for writers and photographers, who would have to contend with the clouds of biting midges called "no-see-ums."

"Almost there," I said once again to Linda, who was overturning crabs, so lagging behind. Near the mouth of the creek, projecting out into the bay, was a sod bank that the birds had been using at low tide. Birds were still there: a couple of dozen turnstones, and a handful of knots. It wasn't the throngs of birds we'd been hoping for, but at least the trip wouldn't be a total bust.

I reached the end of the beach. Edged myself around a tall stand of reed grass, and ...

There, clustered about the sandy bar, were about a thousand shorebirds, all busily foraging for eggs. Lots of semipalmated sandpipers, lots of sanderlings, lots of ruddy turnstones, and best of all, lots of red knots—not the 6,000 that were reported to be up in the bay (about one third of the current estimated populations). But six hundred at least—most of them decked out in their breeding plumage, the silver backs and ruddy breasts that were the source of the old market gunners' name for the bird, "robin snipe."

Linda turned the corner. Stared. Smiled. Here, on a spit of sand some forty feet wide and two hundred feet long, was a concentration of birds that harked back to the wonder and bounty that greeted visitors when shorebird numbers were at their peak.

Linda slipped out of her pack. Cracked the cover. Starting fitting lenses to cameras.

"I'll open the tripod," I offered.

"Thanks," she said.

"Are you going to stand or kneel?" I asked.

"Kneel," she said. "Crawl. I want to be eye level with these birds. I want to taste horseshoe crab eggs. Do you want to shoot?"

"You might have to dig," I said, suddenly conscious of the fact that the birds were not foraging on exposed eggs. All of them seemed to be probing and rooting. It was an adaptation I'd never seen. Back in the days of plenty, only the turnstones were egg mass excavators. Now, it seemed, all the successful birds were doing it.

"I'll leave you the two-hundred-to-four-hundred lens," Linda said, not waiting for my reply. "It's set for program, and you've got a new memory card."

"This is so awesome," she said, lifting her camera fitted with her favorite 500-millimeter lens (named "Big Bertha"). "This is so cool," she assessed. Then she waded slowly into the birds, whose ranks parted to receive her. In less than five minutes, Linda was seated in the sand and surrounded by birds. A 50-millimeter lens would have served her better.

It *was* awesome. It was affirming. In fact, it was perfect. Almost perfect.

UNDOING THE KNOT

I didn't start shooting immediately. And I didn't take notes, which, being the scribe in this endeavor, I should have been doing. What I did instead was savor. The spectacle. The day. The miracle that is a shorebird. The study in pure, applied hard-shelled tenacity that is a horseshoe crab.

The red knot, owing to the rapid and dramatic decline in its numbers, has been studied and written about extensively. In less than two decades' time, the population of the *rufa* subspecies, the subspecies whose strategy for survival seems tied to Delaware Bay, went from an estimated high of 160,000 to about 18,000. In cold, clinical mathematics, its population is now just over a tenth of what it was when I first set eyes on the spectacle, over thirty years ago.

The undoing of the phenomenon was unnecessary and preventable. But, as so often happens when our species ascribes value to another species, it happened anyway.

What happened? The overexploitation of a natural resource—one of the oldest stories in a long, sad book that is the history of our species and the environment.

Beginning in the mid-1980s, horseshoe crabs were harvested as bait. First, by New England fishermen who would drive down to Delaware Bay beaches like this one in tractor-trailers, load up on crabs, drive back to New England, and sell their catch for lobster bait.

Later, in response to a burgeoning Asian market for

conchs and eels, local watermen got into the game and started harvesting the animals, first by hand, later with off-shore dredges working just off the beach. Halved or quartered, the crabs were put in pots and used for bait. The larger females were, of course, favored.

At market, watermen were getting a dollar apiece for female crabs. Good money on the economically strapped bayshore.

Ten to twenty million is a lot of horseshoe crabs, but with up to a million animals being harvested from the breeding population every year, it didn't take long for the population to falter. Female crabs lay a plentitude of eggs—nearly four thousand at a sitting—and a female can breed as many as twenty-five times in a season.

But only a fraction of the eggs deposited in the sand hatch. Of the larvae that emerge, only about one out of thirty thousand lives to celebrate a birthday. They move into deeper water as they develop, taking another eight to eleven perilous years to attain sexual maturity. Then, lured landward by the tug of the moon and the spurs of procreation kicking their carapaced flanks, the adult crabs return once again to the beaches of Delaware Bay to spawn.

Being a mature, sexually active horseshoe crab constitutes one of the planet's greatest single accomplishments, but insofar as *Limulus polyphemus* has been on this planet for about 340 million years and inasmuch as the crabs Linda and I were righting on the beach were the products of a long line of superachievers, if it ever comes down to a contest between my species' survival and the crabs', my money is on the crabs (not that I'll be around to collect it).

Things are different for red knots and other long-distance shorebird migrants. Their winning streak goes back only a couple of million years (if that) and their strategy for survival is riskier than that of the crab, predicated as it is on long, nonstop migratory jumps with the promise that there will be a surfeit of food waiting at the end of that flight, allowing the birds to refuel quickly and reach Arctic breeding grounds in time and in shape to breed. Red knots winter in extreme southern South America—about as far south as a migrating bird can go. They breed on the islands of the Canadian Arctic Archipelago—about as far north as a breeding bird can go. Delaware Bay, on the red knots' current flight plan, is the refueling point on the northbound trip.

When knots reach Delaware Bay, after having spent many hours in the air, they are literally skin and bones. All of their winter-stored fat, and some of their muscle tissue, too, has been burned to get them to their midflight fueling depot.

Refueling takes about two weeks given optimal feeding conditions. If successful, birds about double their weight. This was how it was when crab populations were high and free-floating eggs were there for the picking. Now, with crab numbers depleted and eggs hard to come by, the starved birds cannot make flight weight. They aren't reaching the Arctic in time or in condition to breed. Given poor breeding success and normal attrition, the numbers of knots and other crab-dependent shorebirds have, like the crabs that supported them, dwindled.

It's still a spectacle. It's still worth seeing. And it is probable, given a moratorium on the harvest of crabs in the bay, that the crabs and the birds will recover to preharvest levels. Some experts believe a significant recovery is no more than a decade away.

So it might well be that here on Thompsons Beach on the twenty-fourth of May 2008 what Linda and I were experiencing wasn't a window to the past. It might be that what we were seeing was the future.

Almost Perfect

The wind held for most of the morning, making the temperature ideal and keeping the hosts of biting midges at bay. No sky shines quite so blue as a cold-front blue. The Arctic air that spawns these dry, fair-weather systems seems to whet the sunlight and put an edge to all it touches.

Likewise, there is no green like salt marsh green. It is deep and rich and pure; untainted by blue, untinged by yellow. Just pure, pure green. Green enough to make the Emerald Isles want to trade up. Green enough to make you wonder what the rest of Eden was like, because after its equilibrium was shattered by a simple act of harvest, it is pretty clear that some of it washed up here, on the shore of Delaware Bay.

I heard the sound of the motor long before the small commercial boat rounded the point of land. A sixteen-footer, with barrels to hold the drift net in the bow and a capped and handkerchief-hooded figure in the stern. Working the tide, just as we were, he turned into the wind and started hauling net by hand. His was the only commercial boat we'd seen.

I recognized the fisherman (his makeshift legionnaire headdress was unmistakable) but couldn't say I knew him. His name was George Kumor, "Captain George," as he is universally known and addressed. He fishes the waters off Thompsons Beach, putting in on every tide at the boat ramp by the gate.

George was a curious mix of talkative and taciturn. Straight talking, critical of environmentalists and regulators, but candid about his commercial colleagues, too.

In one sense, we were adversaries. He a waterman, I an environmentalist, both competing for the same resources— the environmental riches of Delaware Bay. But we were allies, too. Both of us wanting, ultimately, the same thing: a healthy and environmentally rich region.

Because that is what we need to sustain us, and that is why we live here.

No, I didn't know George. Not yet. But the story of the bayshore is the story of people and nature, and how each has molded the character of the other. Telling his story is as much a focus of this book as is extolling the virtues of the region. At that moment I knew, or at least hoped, that George's story was going to figure in this one.

You better hurry, I thought to the man hauling net. You're losing your tide. But the caution was wasted, because when it comes to knowing the water, there is nobody better than a bayshore waterman. Chances were good that the bayman was thinking, you better hurry whatever it is you are doing, Mr. Bird Watcher, because the wind is dropping and in about a Thompsons Beach minute you are going to be sitting in a cloud of no-see-ums.

And he was right. But before that happened, I had just enough time to lean back, sweep my eyes across the unencumbered landscape, inhale the good, salty tang of the marsh, feel the bite of sand on bare elbows, and take stock of this moment in time.

I was on a beach with the person I love, who, sensing my study of her, turned and smiled—the smile of a happy person doing what she loves most to do. I was surrounded by a miracle whose magic vaults ages and hemispheres. It was as sunny as a summer day, as cool as a morning in spring, and there was a breeze in my face. Except for the lone waterman, now heading toward the mouth of the inlet, there was not the hint of any other soul.

It was Memorial Day weekend. We were alone on a beach in the most crowded state in the Union. The world was perfect and we were complete.

Somewhere there were people sitting in traffic, knowing that they were going to be late for the barbecue.

But it wasn't here.

Somewhere someone was engaged in a third circuit of Cape May streets, searching for a free parking space.

But it wasn't us.

Somewhere there was a shopper who could find every

size for the shoes she wanted except the size that fit; a homeowner discovering that flipping the circuit breaker for a water pump that hasn't been used since he closed up the shore house last autumn doesn't guarantee water; and a young father and husband just discovering that in the rush to get all the stuff needed into and onto the car, he left one bicycle behind. Hers.

But none of these things troubled the perfection of here, and the reason I know that this moment was perfect was because it ended. The wind suddenly stilled, then turned. The sound of a low-flying single-engine plane rose above the chatter of the sandpipers and the stutter of turnstones, and as the plane drew near, hugging the beach, the birds erupted into flight.

It might have been the shorebird survey team doing their census, or perhaps a plane chartered by a photographer who aspired to get some aerial shots of the birds. The irony that it was those who were interested in the birds, not the commercial fishermen, who were forcing the birds to burn precious fuel, wasn't lost on me.

After climbing and circling, the birds lit on the nearby exposed sod bank, then lifted off again, settling on the far shore of the creek, the side away from us. Thirty seconds later they were up once more, and Linda used the opportunity to rise, gather her gear, and head to where I was waiting.

The birds circled, then settled for good. Though they were initially flushed by the plane, it was nervous anticipation that was keeping them aloft. The birds had also felt the shifting wind, the turning season. Birds heading for the

Arctic like a tail wind to see them home. The wind, coming in off the bay, was fanning their anticipation.

There is a German word to describe the nervous energy exhibited by these birds: *Zugenruhe*. It means "premigratory restlessness."

"Did you get some decent shots?" I asked when Linda drew near.

"Maybe," she said, trying to hide a smile.

"Quit 'cause you filled your memory cards?"

"No. Quit because the light is too harsh and the plane spooked the birds."

I nodded. "Wind shifted, too," I observed, not certain Linda had noticed.

"Felt it," she said. "No-see-ums came out as soon as it dropped."

"I'll bet some of these birds go tonight, maybe all. Need help with that tripod?"

We made it back to the car without losing our boots (or balance). The Wawa was still jamming; in fact, it was busier than ever. Filled with people on their way to someplace else. People who've never heard of Thompsons Beach and never seen the wings of a thousand shorebirds reaching for the sky and maybe never will.

That evening, about an hour before sunset, as Linda was downloading images and I was monitoring two Memorial Day weekend steaks on the grill, I heard the xylophone-strumming stutter of short-billed dowitchers. I looked up and saw a V-shaped flock speed overhead, climbing and calling as they went. Their direction was north. Their destination, the tundra just east of Hudson Bay. The season's last great exodus of northbound migrants had begun.

Within a minute the first group of two hundred birds was followed by another and, before the steaks were done, two more. Next stop, the Arctic, where summer is more a visitor than a resident.

As I lifted the steaks onto a plate, a baby robin chirped from beneath a nearby hedge. The year's first fledgling. One season ends, another begins. If the song of blackpoll warblers sounds Taps for spring, the chirp of newly fledged birds is the fanfare for summer.

Bayshore summer.

Chapter 2
Early Morning, Mauricetown

I woke before the alarm (meaning our noodge of a dog, Raven) and eased myself out of bed so as not to trouble Linda's sleep. Scheduled to work a B shift, she wouldn't have to rise before 5:00 A.M.

What time was it now? Early. Very early. But if you are going out on the water, early is when you rise.

Slipping into laid-out clothes, turning right at the door, I navigated the stairs in the dark (which, given the pre-building-code pitch of stairs found in homes built in the mid-1800s is more challenging than it sounds). Reaching the first floor, I turned hard left and found myself, six paces later, in the kitchen.

They built houses with small rooms in 1861, too.

A stranger who stopped by one day, explaining that

she'd lived her childhood in our house, told us indoor plumbing had arrived in the 1940s—which seems about right. When Linda and I bought the house, in 1990, the plumbing, along with just about everything else, was in need of replacing.

With the mindfulness accorded the historic nature of the structure and town, we kept our modifications to a minimum. Were he to visit the home on Second Street now, the ship's blacksmith, who, according to records, built the structure, would find little to discomfit his memory of the place. It is now, as it was then, a modest home with a generous porch, located in the shadow of the Methodist church steeple that dominates the landscape and the town.

That steeple was the first thing I saw when I stumbled upon Mauricetown in 1979, while conducting a survey of breeding northern harriers (or "chicken hawks" as they were known locally). A tall white spire, marking the location of a storybook town, overlooking a broad tidal river.

Except for the "Wheaton Mansion," owned by one of the kings of the region's glass industry, you could have bought any house in town for under fifty thousand dollars back then. And while I wasn't in a home-buying mood, I did keep the comely little village in mind. So in 1990, married and having traveled all over North America to complete the book *The Feather Quest*, we decided to settle down.

"There's a little town I'd like you to see," I said to Linda soon after discussions got serious. So one Sunday morning we drove to Mauricetown. Nosed our car up and down the quiet streets. Saw a For Sale sign in front of a home, along with a banner that read OPEN HOUSE TODAY.

We walked through the front door. Linda looked left, looked right, looked at me, shrugged, and said: "We'll take this one."

The house wasn't all that important anyway. What we wanted was the town and all the protected, natural lands around it. Take a look at a map of Cumberland County, and you'll see what we saw. Thousands of acres of natural and protected ecofriendly green, and Mauricetown sits in the middle of it. It's not what people think of when they think of New Jersey.

YOU LIVE IN NEW JOISEY?

Grabbing a mug of coffee, I headed for the side door, turned the knob (few people bother to lock their doors here, and I doubt there is a dead bolt in the entire town), and stepped into the night. In the street, one of the many cats that own the town after dark delayed her mission long enough to appraise and dismiss me. High overhead, cloaked in the luminous veil cast by the full moon, male purple martins were filling the air with soft burbles, churps, and trills.

Aside from these prospecting swallows and the caroling of robins (who, tricked by the light of the moon, had treated residents to their "dawn chorus" all night), the town was silent. Elsewhere, on commuter approaches to New York and Philadelphia, traffic was building. Here in the village, we were as far from mayhem as a person can be and still send tax dollars to Trenton. Trenton, New Jersey. Arguably the most maligned state in the Union.

We get it all the time, of course: questions from folks who want to know why two well-traveled people who love the natural world live in this most densely populated state. People are mostly astonished when we tell them that we live in New Jersey by choice.

If this assertion is met with skepticism (signaled with raised eyebrows, a smirk, and a reference to the view from the New Jersey Turnpike), I sometimes add, "Yes, we spend tens of millions of dollars every year making the view from the turnpike look so ugly. It keeps people [like you] from wanting to move here."

The "like you" is usually just implied.

But most people respond with genuine puzzlement. To them I offer this:

"Were I to airdrop you, blindfolded, into the place I live and invite you to tell me where you are, you wouldn't guess New Jersey before you were almost out of states."

While this is not an exaggeration, it is only half true.

Which half?

The southern half. I was born in the other half, in Morris County, to be precise. A place that was, in my lifetime, transformed from a rural corner of America to a suburban labyrinth when it seemed half the Fortune 500 companies in the United States decided to relocate their corporate headquarters to North Jersey.

The resulting development and congestion undercut their reason for moving there.

In the fifth grade, in accordance with the state's grade school curriculum, I was treated to a year of state-favoring study. In short sum, and in case you are not the product of a New Jersey public school education, it went something like this.

New Jersey, the fifth smallest state, is a geographic peninsula surrounded by water on three sides. Its 7,787 square miles are bordered by the Hudson River and Atlantic Ocean to the east, Delaware Bay to the south, and the Delaware River to the west. The fifty-mile northern border it shares with New York is all that keeps it from being an island.

New Jersey is one hundred and sixty-six miles long, fifty-seven miles at its widest point, and its boundaries embrace five landforms. From north to south these are the Appalachian Ridge and Valley Region, the Highlands, the Piedmont, the Inner Coastal Plain, and the Outer Coastal Plain, the last expanse falling roughly south of a line drawn from Sandy Hook to Salem, encompassing 41 percent of the state.

If you remembered all of this, you got a 100 and Mrs. Reynolds stuck a gold star on the top of your test.

Cumberland County lies wholly within, and at the southern reaches of, the Outer Coastal Plain. Only Cape May, which in geographic and vegetative character resembles coastal Cumberland County, lies farther south. Mauricetown, where Linda and I live, lies at the same latitude as Baltimore, Maryland. Cape May (Baja Cumberland) is due east of Washington, D.C.

That's right. Much of South Jersey and all of Cumberland County falls below the Mason-Dixon Line.

As the term *coastal plain* implies, South Jersey enjoys

both semantic and geologic kinship with that large body of water lying east of New Jersey, the Atlantic Ocean. In fact, at times, they have been one and the same. Over the course of many millions of years, New Jersey's coastline has expanded and contracted in response to rising and falling sea levels. There have been times when the southern boundary of the state followed, roughly, a line drawn from Staten Island, New York, to New Jersey's capital, Trenton (and what we now call both the inner and outer coastal plains were completely submerged). At other times, when much of the earth's water was locked up in ice, the "Jersey Shore" extended all the way out to the edge of the continental shelf—about one hundred miles off New Jersey's present coastline.

The impact on Memorial Day commuters would have been profound.

These periodic submergences shaped the geographic character of South Jersey, making it low and mostly flat with a gradual gradient to the east. The highest point in Cumberland County is 140 feet above sea level and lies in well-named Upper Deerfield, the northernmost township. The lowest point is sea level, or bay level insofar as Cumberland County does not abut the Atlantic. Its southern border is Delaware Bay—New Jersey's "west coast."

Given South Jersey's long, mercurial relationship with the sea, you might guess that the substrate of the coastal plain is mostly sand—and you would be right. South Jersey rests upon hundreds of feet of sand. The source was several mountain ranges that sprang up, and wore down, in the northern and western portions of the state. The granular refuse was ferried, via assorted watercourses, to the sea, and in a very real sense, South Jersey is just a reconstituted, redistributed bit of North Jersey and nearby portions of Pennsylvania and New York.

Much of the Outer Coastal Plain is composed of coarse quartzite sand—a soil type called by geologists the "Evesboro-Klej-Lakewood" association. It is poor in nutrients and porous to a percolating fault. It is the soil type that underlies much of New Jersey's pine barrens.

The Jersey Devil, whose name is promoted every time New Jersey's professional hockey team takes the ice, is a mythical creature whose cloven-hoofed tracks lead straight out of the pine barrens. The blueberries you buy in the produce sections of supermarkets in Fort Worth, Texas, and Bozeman, Montana, in July were very probably grown and harvested in the barrens.

But berries are among the few cultivated crops that do thrive in the region's sandy soil. The name "barrens" almost certainly applies to the land's agricultural potential, which is, forgetting for the moment a wealth of scrub oak and pitch pine, that in the days before petroleum-based briquets supported a thriving charcoal industry, about nil.

The pine barrens extend south into Cumberland County, defining much of its eastern border. Soils in northern and western Cumberland County are immensely richer and more supportive of agriculture, and farms dominate the region. If you have ever wondered why New Jersey is called the "Garden State," take a drive through Upper Deerfield or Fairfield or Lawrence townships anytime from April through September.

Linda and I eat so much locally grown asparagus in April that we are close to suffering kidney failure by May. Tomatoes are a New Jersey resident's birthright. White corn, boiled to sweet, tender perfection, is considered an art form.

Bordering the coastal marshes are the Hammonton-Fallsington-Pocomoke and Huck-Atsion-Berryland soil types. Supporting, for the most part, forests whose vegetative character harks back both to the pine barrens to the east and to the Delmarva Peninsula to the west, the soils are particularly significant to the culture and economy of the region, not for what they are but for what they overlie, which is high-grade sand, laid down by ancient seas and now commercially mined. It's not for nothing that, in the 1800s, Millville flourished as a glass manufacturing center and that sand mining still reigns among the region's dominant industries. When I first came to South Jersey, stores in towns such as Port Norris and Dividing Creek wouldn't even sell soft drinks in cans.

There is one final habitat type whose qualities impart their character upon the region. This is tidal wetlands. These constitute the county's southern border and are a world that has the character of both land and sea. Twice each day these low-lying coastal areas are claimed by the tide. Twice each day, the water retreats and their allegiance switches to the land.

This yin and yang not only gives balance to the region but nurtures it. Tidal marshes rank among the most energetically productive ecosystems on the planet—engaging in a rich, green commodity exchange in which matter and energy are forged into biological currency, temporarily banked in plant tissue, then traded up to higher trophic levels, or cycled back once more into the tidal exchange.

Approximately 28 percent of Cumberland County's 677 square miles are tidal wetlands. And while I have the entire county to roam about in, this is where I spend 90 percent of my free time. There are multiple reasons, but two are determining:

Reason 1: I am here.

Reason 2: The rest of the planet's inhabitants are not.

DAWN'S EARLY LIGHT

Over the ascending sound of early-rising birds, I heard the measured patter of footsteps and caught the form of Lisa McDermott, the town runner, turning the corner and heading in the direction of her husband, two young sons, and home on the south edge of town. Lisa and her husband, John, originally from New York State, are, like Linda and me, among the growing number of transplants to the region. Across the street is our neighbor Bill Hoover, an eighth-generation bayshore resident (and high school track coach), and just around the corner is Irene Ferguson, a lifelong resident and 1937 graduate of the town's two-room schoolhouse.

Together this amalgam of old and new comes to about

three hundred inhabitants apportioned among about one hundred homes—just about the population of the town in 1932. In its heyday, in the 1880s, there were seven hundred residents—enough to support a Baptist and a Methodist church, six general stores, a canning factory, two shoe repair shops, a hotel, a funeral home, and a post office.

Only the Methodist church and post office remain. These enterprises, in addition to a dozen or so antiques shops, constitute the town's businesses. The social and cultural center is the fire hall. Millville/Vineland, the region's burgeoning shopping center, is twenty minutes' drive. Philadelphia is forty-five minutes (if there's no traffic) and the Philly airport an hour from our door.

I took another sip of coffee and looked at my watch, which confirmed what the eastern sky was telling me. It was time to go.

Draining the mug, I left a note to Linda. Got in the car and headed west, navigating a course around late-prowling cats and early-rising joggers.

Elsewhere in New Jersey, there were people racing for planes, buses, and commuter trains. I had a boat to catch.

CHAPTER 3
The Waterman's Blues

I reached the established meeting place by 5:20, chuckling at my initial concern about not finding the right spot: "the two-story house with lots of crab pots in the back." Among the cluster of houses that constitute the hamlet of Money Island, there is only one two-story house. Ten minutes later, a pickup truck of indeterminable color pulled up, and without much discussion the driver invited me to get in. I did, taking all that was left of the space in the front seat, already occupied by Captain Tom Pew and Mate John Burens.

It was a workday, and both men were eager to be on the water (not that there was anything remarkable about this). Watermen with nets to haul and traps to pull are on the water every day in season. Only bad weather keeps them tied to the dock, and today's weather was perfect. Warm, sunny, light winds from the southwest.

A great day to be on the bay.

A hot orange ball of a sun was peeking over the horizon by the time we reached Tom's workplace, a thirty-five-foot fiberglass head boat powered by a diesel engine that seemed not to have even the suggestion of a muffler; fitted with a power winch and wheels fore and midship. No brass, no polished wood, not even a name. It was a working boat, about Tom's eleventh, as near as he can recollect.

Tom is forty-one, handsome, in a rugged sort of way, and built like a halfback—not tall but solid. When he is not laughing (and he laughs often), his face is most commonly creased by a smile.

He started fishing on his father's boat when he was fifteen. When pressed for his ancestry, he can trace it back to a great-uncle who was a trapper and bayman, and quite obviously something of a role model, but beyond this Tom's recollections get vague. When he is asked about family and children, his memory is more precise, but the players are more varied and confusing, involving multiple spouses and both natural and adopted children.

His current wife, Trish, a registered nurse, is the owner of the two-story house, which she bought from her parents. Their son, Tommy Jr., just got his own eighteen-foot boat and plans to make a living on the water like his dad.

John, Tom's mate, is taller, shyer, a cigar-store Indian of a man. Dark-haired, aquiline-featured, he wears an expression that hesitates between stoic and bemused. A resident of Cedarville, he goes fishing even in his spare time, in fact, every day after work.

Both men smoke. John admits to three packs a day. Both men dress in jeans and T-shirts preanointed by a soap-defeating grime. After loading the boat, they suit themselves up in work-stained, bibbed rain pants—Tom in white Grundens, John in rain slicker yellow Carhartts. Tom wears a camo-patterned baseball cap; John goes bareheaded. Neither man uses sunglasses. Neither man bothers with sunscreen.

Why would a man who smokes three packs of cigarettes a day worry about skin cancer?

"Why don't you sit on that cooler," Tom directed as he stepped into the wheelhouse and John slipped the lines. It was about the last decipherable sentence I heard for the next six hours. Moments later, Tom fired the engine, cranked up the CD, and ramped up the volume on the marine-band radio to decibel-trumping levels. Five minutes later we were at the mouth of Nantuxent Creek and Tom really opened her up.

It was only when we stopped at the first string of crab traps and Tom throttled back the engine that I realized the song coming out of the speaker near my left ear was the Rolling Stones' "Brown Sugar." As musical accompaniment goes, not particularly apt. Most of the watermen working the bay are white and male. However the water, this high up in the bay, does have a distinctly brown quality. Nurtured and freshened by the Delaware River, it is an ideal habitat for the blue crab. The animal that is, now, the commercial fishing mainstay of local watermen.

Tom didn't anchor. He aligned the boat so that the floating marker buoys passed on the right side within reach of his six-foot gaff. Playing the balancing forces of wind, tide, and diesel against one another, he nudged the boat forward, steering less by hand than by will and an innate understanding of the water, just as watermen have navigated the waters of Delaware Bay for nearly four centuries and Native Americans did before that.

HISTORY LESSON

The history of New Jersey's bayshore is less the story of people and the land than it is of people and the water. True, communities sprang up along inland creeks and rivers that offered ready access and a measure of buffering protection. But what fortunes were to be made, particularly in the early period of settlement, people either drew from the water or ferried atop it.

It was only after World War II that the economic and social importance of this great body of water, second in size and volume, as bays go, only to Chesapeake Bay to the south, faltered. But the imprint the water made on the people and the region's culture persists. So while watermen like Tom and John might now constitute a numeric minority, their tradition goes back a long way, and when you look at men like these, you understand better the ruggedness and resourcefulness of people who live here.

When the first European explorers arrived, they found a

native people already well established and, judging by the length of their tenancy, eminently pleased with their corner of the planet. The Lenape, literally "common man," are a tribal group linguistically linked to the Algonquian people. Historians believe they migrated north out of northeastern Virginia and the Carolinas about 4000 B.C., settling in northern Delaware, eastern Pennsylvania, and all of New Jersey. Their population during the time of European exploration was estimated to be about twelve thousand.

In Cumberland County, there were six permanent Lenape communities numbering about 250 souls each and scores of smaller encampments. In the summer, bands living away from the bay relocated to set up seasonal camps, while those living within five miles of the coast made frequent trips, setting up temporary camps where freshwater streams entered the bay.

Buried trash pits, or middens, mark the locations of these summer encampments. These refuse pits contain large numbers of turtle shells, as well as the bones of waterfowl, turkey and deer, and other game.

But if the volume of clam and oyster shells means anything, it seems clear that the native people came here mostly for the seafood and that the wealth of it was enough to offset the plague of biting insects that could hardly have been any less troublesome than they are today.

Like other eastern tribes, the Lenape fared poorly in the face of European encroachment. Not particularly warlike, uncomprehending of the European concept of landownership, the people were gradually and resentfully forced west into Pennsylvania, Ohio, and Indiana. Perhaps because of their disgruntlement, the tribe allied itself with the French in the French and Indian Wars, a decision that did little to restore their fortunes. For what historic significance it merits, in 1778 the Lenape were the first Indian tribal group to be recognized by the incipient government of the United States.

The ancestors of the Lenape are now found in Oklahoma, far from the waters of the bay. Their geographic estrangement notwithstanding, they are now known as the "Delaware," like the bay. Unlike the names of local waterways, such as Manumuskin and Nantuxent, the name "Delaware" has no linguistic ties to the Lenape. The bay was named for Thomas West III, the English Lord De La Warre.

It was on August 28, 1609, that Henry Hudson, sailing for the Dutch East India Company aboard *De Halve Maen* (or the *Half Moon*), anchored in Delaware Bay, establishing Dutch claim to the land on either side. His captain and fellow countryman Cornelius Jacobsen Mey reinforced that claim by sailing all the way up the bay to the mouth of the river four years later. Settlement followed.

In the early to mid-1600s, groups of Dutch, Swedish, and English settlers arrived in the area and almost immediately began vying for economic, numeric, and military supremacy. The Swedes, who may initially have enjoyed a short-lived numeric edge, established settlements near Wilmington, Delaware, in 1635.

In 1643, in response to mounting pressure from the Dutch, they constructed Fort Elfsborg across the bay in Salem County, New Jersey. The fort was abandoned a decade later, the garrison defeated not by force of arms but by the hordes of mosquitoes that made living there intolerable.

In 1644, the English won title to the New World holdings of the Dutch, including the land surrounding Delaware Bay. It was a package deal. They acquired New Amsterdam (now called New York City) at the same time.

In 1664, King Charles II gave to Lord John Berkeley and Sir George Carteret the colony of Nova Caesarea (New Jersey), although some accounts say that Charles first gave the land to the Duke of York, who then sold it to Berkeley and Carteret. Thirteen years later, a newly converted Quaker and commercial loose cannon named John Fenwick purchased Carteret's holdings for the sum of one thousand pounds sterling. Then things began to get deliciously litigious.

Another Quaker, Edward Byllings, claimed that Fenwick had made the purchase using his, Byllings's, money. Untroubled by this, or perhaps deeply troubled and looking for escape, Fenwick and a group of fellow Quakers sailed to the New World and landed near Salem, New Jersey, on September 23, 1675. It was there, as legend has it, beneath the spreading limbs of the massive and still-standing Salem Oak, that Fenwick negotiated with thirteen Lenape chiefs for the hunting and occupancy rights to the land in what would become Salem and Cumberland counties. This settlement did nothing to quell the claims of Byllings, who was by that time bankrupt, nor Byllings's creditors, so Fenwick petitioned his fellow Quaker William Penn to arbitrate the matter.

Penn met with Carteret to certify Fenwick's claim. The resulting Quintipartite Agreement divided Nova Caesarea into West Jersey and East Jersey—basically splitting the state, north and south, along a line running from just north of Atlantic City to extreme northwestern New Jersey. It was further decided that Fenwick was entitled to only one tenth of West Jersey. Billy Penn, as he is affectionately known and whose statue stands atop City Hall in Philadelphia, awarded the balance to Byllings and his creditors.

Fenwick's difficulties didn't end there. In 1682, William Penn obliged him to surrender all but six thousand acres in the vicinity of present-day Mannington Marsh, Salem County, as payment for legal services rendered. Fenwick ultimately relieved himself of his legal difficulties by dying—but not before he'd embarked on another land scheme involving plans for a town he called Cohansey, after the river upon whose west bank it would lie. The town is now known as Greenwich and, in defiance of pronunciations accorded the name in New York and Connecticut, it is really pronounced *Green witch*.

If you want to brand yourself an outsider in Cumberland County, ask a native directions to "Grenitch." Don't be surprised if half an hour later you find yourself on the New Jersey Turnpike, heading north.

In 1702, the colonies of West and East Jersey were reunited. In 1784, in the interest of making local government more accessible to those living in the distant, eastern half of Salem County, the county was partitioned. The new, eastern county was named Cumberland County after the Duke of

Cumberland, the popular man of the hour, whose defeat of Bonnie Prince Charles ended Stuart claims to the throne of England.

Surprisingly, the location chosen for the new county's governing seat was not the beautiful and thriving seaport of Greenwich. The honor fell to a strategic cluster of homes about ten river miles upstream called Cohansey Bridge (now Bridgeton).

The land disputes resulting from Fenwick's transactions and scheming outlived him, and disputes over boundaries and land claims continued to plague authorities for many years. So if you ever wondered why it is that New Jersey is the most litigious state in the Union, you now have part of your answer: it started that way.

DANCE STEP

"The shit gets to slingin' now," Tom announced as he darted from the wheelhouse, my only warning that it was time to get out of the way. Tom, standing at the port rail, jabbed at the water with his gaff and lifted his prize, a buoy. All the buoys are numbered. The number that corresponds to the one on Tom's commercial fishing license. In a seamless sequence that takes longer to describe than to effect, he affixed the float line to the wheel of the winch, which in just under ten seconds drew the square, cooler-size wire trap from the water, and he passed it to John's waiting hands.

As Tom then guided the boat on to the next float, about a hundred feet away, John gave the trap a hard shake or two or three, emptying the contents into a stainless steel trough.

Then he rebaited the trap with a whole bunker and dropped it overboard. Then, in a precision-directed frenzy, he went about the serious business of sorting through the catch—a process that consumed almost exactly thirty seconds. Just enough time to reach out and grab the next trap that Tom had hauled and was offering forward.

Fifteen seconds to hook and haul.

Fifteen seconds to shake, rebait, and return.

Less than thirty seconds to sort.

Do it again.

There are fifty traps in a string. It takes about fifty minutes to check a string of traps if everything goes like clockwork—and it does. The coordination between men and boat appears seamless, a dance involving three partners, with a single repeated series of steps.

It was mesmerizing. It was also hard work. Even empty, the weighted wire-cage traps weigh between ten and twenty-five pounds. The catch, by-catch, and sodden algae that fouls the wire may double or triple a trap's weight. There is not a lot of body fat on Tom or John.

There were also, I noted, beneath the sheath of orange rubber gloves that both men wore, five fingers to every hand, and a notable absence of scars on their tanned and well-tuned forearms.

They knew their work. They were very good at it. And the proof was piling up in the peach baskets arrayed at John's feet. The creature that almost single-handedly, or clawedly, puts the commerce in commercial fishing in Delaware Bay.

The Atlantic blue crab is found along the western edge of the Atlantic Ocean, ranging from Nova Scotia south to Argentina. Made famous by James Michener's novel *Chesapeake* and wonderfully described in William W. Warner's best-selling tribute *Beautiful Swimmers,* the crustacean makes a claim to this praiseful tribute borne out by its scientific name, *Callinectes sapidus. Callinectes* is Greek for "beautiful swimmer." *Sapidus,* in Latin, means "savory," and while we may appreciate this crustacean's ability to move swiftly and gracefully in water, we prize it for its epicurean quality.

Search the world over, you'll find few crabs whose succulent white meat is quite so savored and prized as that of the blue crab. You have to earn it. The armored morsels are small relative to their larger and more finger-picking-friendly relatives the Dungeness crab or Alaskan king crab.

I used to colead a birding tour for Victor Emanuel Nature Tours that took in Hawk Mountain, Pennsylvania; Cape May, New Jersey; and Bombay Hook National Wildlife Refuge in Delaware. The tour included lunch at a waterfront eatery in Leipsic, Delaware, called Sambo's—a place celebrated for its steamed crabs, served in the traditional fashion, on newspaper-covered tables, and consumed, by novice and experienced crab lovers alike, with carapace-ripping, claw-cracking, leg-sucking gusto. It was a great tour, offering world-class birding opportunities. But tour evaluations showed time and again that the thing participants savored most was lunch at Sambo's.

In Cape May, there are restaurants that brag that their recipe for crab cakes is true to the one detailed by Michener in *Chesapeake*. The crabs go for $30 a plate in better restaurants, or live for about $100 a bushel retail for the larger grade of males at 2008 prices. Watermen, of course, see about half this amount for the animals they catch, with the largest male crabs reaping $65 a bushel and smaller males and females garnering $35. By Tom's estimate, it costs him $250 in expenses every time he goes out on the water. What makes the difference between an economically successful run and one that loses money is the volume and quality of the catch.

It's not predictable. A day's catch is influenced by all manner of variables—tide, moon, date, spawning success, population numbers, trap placement, competition, and, of course, state regulations.

Every time a crabber goes out, he throws dice with the universe. It is at once commercial fishing's greatest risk and greatest attraction. And nobody plays this game who is not optimistic and who doesn't like to play hard.

"I think we're going to catch a ton of crabs today," Tom said aloud, to the universe and to John.

"Yep," said the cigar-store Indian. The universe, predictably, said nothing.

But is the beautiful swimmer beautiful?

This probably is not a fair question, since the focus of

the genus name stems from its swimming ability, not its physical appearance, and boy can this animal swim. Still, I couldn't help but wonder as I watched John sort by size, sex, and stage of molt.

Males five and a half inches wide or wider in one basket (number 1s). Smaller males, at least five inches wide, in another (number 2s). Females, at least four and a half inches wide and without the spongy egg cases attached to their abdomens, in their own basket. "Lights," or recently molted animals whose carapaces are not yet hardened, in another. "Shedders," or softshell crabs, stored separately.

Shedders, sautéed in butter and eaten whole, are a delicacy.

But at question is the animals' visual, not gastronomic appeal, and, as crabs go, the blue crab is indeed a designer crustacean, built for show *and* speed. In size and shape it resembles a flattened, multilegged croissant, but a croissant armed for battle. The body is encased in a light, hard carapace or shell—essentially a skeleton it wears protectively on the outside. The upper parts are green-brown or weathered bronze; the underparts, particularly in freshly shed crabs, alabaster white.

The segmented appendages, five to a side, are likewise dark above, light below, but tinged at their bases with turquoise, the trait that gives the animal its name. The hind limbs, which are mostly blue, are flattened distally to serve as swimming fins. The forelimbs, or claws, are touched with blue and, especially on the females, red.

There is no mistaking the purpose of these pincerlike instruments. Being nipped by a blue crab is a rite of passage for all young crabbers and the principal reason John was wearing gloves.

On the muddy bottom or swimming in the darkened waters of Delaware Bay, the animals are almost invisible. After eight or ten minutes steaming in a pot, they turn bright, bright red.

The signal for the feast to begin.

But first they have to be caught. And few enough of the ones that are caught can be kept.

The deftness of John's movements was matched and guided by his eye. His intensity was evident, his eye and aim almost unfailing.

In a moment that passed quicker than thought, he would sex and size each individual extracted from a trap, throwing each keeper into the proper basket with a determined fling (to keep the crabs from latching on to his gloves and slowing the process). Females bearing eggs and undersize crabs were tossed, Frisbee-fashion, overboard. These constituted the majority of the catch.

As I watched, I kept track, noting the ratio of crabs kept and tossed per trap.

Two keepers, eighteen tossed.

Six keepers, eleven tossed.

Five keepers, nineteen tossed.

Four keepers, thirteen tossed.

On average about one in every five crabs makes the

grade. The biggest catch of the day was thirty crabs (of which four were keepers). On multiple occasions, the traps came up empty.

At the end of the first string of traps, there was almost a full basket of number 2s; a third of a basket of she crabs, and a handful of the prized, number 1 jumbos.

It was a fair catch, but not one living up to Tom's expressed hopes of "a ton of crabs." If it continued this way, Tom figured they might break even for the day.

Tom retired to the wheelhouse to chat on the CB and have a smoke. John refilled the bait cooler for the next string of traps. It was 7:35. Ten minutes to the next set of pots, lying just off the beach of Fortescue.

BAND OF CRABBERS

Commercial crabbing is strictly limited. In New Jersey there are a fixed and finite number of permits, and these are jealously guarded, passed on within families. But at any time in recent years, there have been probably fewer than one hundred watermen working the New Jersey side of Delaware Bay.

Watermen are, by nature, an independent lot, and most were born to the water, that is, come from a fishing tradition. Many are part-time, augmenting day jobs with their commercial catch during the summer months. Only a few are like Tom, full-time, year-round, fishing for whatever nature provides and whatever they can legally catch: Perch when the commercial season opens in February. Shad later in the spring. Crabs and bunkers in the summer. Bluefish in the fall. Conchs and eels in the off-season. Weakfish, when the regulations (and numbers) provide.

"I do it all," Tom said, lightly. "Got to," he added, not so lightly.

He'll stay out on the water until December. Spend the colder months working on his boat and getting his gear ready for the next season. Go out again when the season re-opens in the spring. Every day he is on the water he is required to record his catch. Every month he must file a report with the New Jersey Bureau of Marine Fisheries. Failure to do so courts the risk of losing his license, thus his livelihood.

The fishermen compete with one another, of course. Compete for the resource, compete for prime places to drop pots and string nets. By and large, it's first come, first served.

The season's catch limits are determined by the Marine Fisheries Council, a body of nine members appointed by the governor. Commercial fishermen constitute a five-to-four majority, and, because of this, decisions tend to be support-ive of commercial fishing interests, giving fishermen, whose livelihood depends upon their catch, the benefit of the doubt. At one time, there was also a Waterman's Associa-tion, which was supposed to formulate and represent the interests of commercial fishermen among competing inter-ests, but, according to Tom, "It sort of fell apart."

Which is too bad, from the commercial fishing perspec-tive, because more and more, as the numbers and types of marketable fish decline and as competing or impacting interests increase, the interests of commercial fishermen get drowned out.

Competitors include sport fishermen and the environmental community, who, in the spring of 2008, won a legislative battle to continue a ban on the harvest of horseshoe crabs in New Jersey waters.

"Don't even get me started on them crabs," Tom said to me with a tightlipped grimace.

And there was something about the way he said it that made me think Tom was well aware that I was "one of them," a member of the environmental community—in fact, an employee of the organization that had spearheaded the moratorium.

"All I know," said Tom, "is when things go wrong, everyone always points the finger at us."

"How many of those finger pointers do you think eat fish?" I asked.

"Most of 'em," he said. "Don't you?"

I do indeed.

We reached the second set of pots. Tom and John repeated the process, taking about as many crabs as they had before. Filling and capping one box of number 2s, and starting another. But the catch in the next set of pots was disappointing, and the fourth set, off Bay Point, less fruitful still.

"What the hell happened?" Tom asked of an unsympathetic universe, and the universe, as if to underscore its dispassionate regard, was even more parsimonious with the day's fifth and final run of traps. For emphasis, the last three traps came up empty.

Tom, not smiling now, went back into the cabin for a smoke and to confer with fellow watermen on the radio. John put the lids on the day's haul of crabs and wrote numbers and volumes on them.

The day's efforts showed two bushels of number 2s; three fourths she crabs, one fourth number 1s. Doing the math, based on current prices, the catch, at market, would gross Tom approximately $129.25. Still a long way from breaking even.

"I'm hoping the bunker will pull us out," Tom said, optimistically. With that, he set a course to intercept the first of his two gill nets.

THE LESSONS AND LEGACY

The first commercial fishery to flourish in the waters of Delaware Bay was the whale industry that began in the early 1600s with whalers coming down from New England, seasonally, to hunt and process the large numbers of right whales that gathered, also seasonally, in the bay.

It was also the first commercial industry to collapse. By the late 1700s, whales were commercially extinct in the bay. The whaling industry, which was just entering its heyday, moved on, in ships out of New England, to other whale-rich waters around the globe—an era immortalized by the American classic *Moby-Dick*.

The North Atlantic right whale, a slow-moving creature that could be spotted from shore and intercepted and killed with longboats, was hunted not quite to extinction. But its population has never recovered, despite decades of protection. Considered a globally endangered species, the current world population of North Atlantic right whales is believed to number fewer than three hundred individuals.

Another creature harvested to economic exhaustion in the bay was the Atlantic sturgeon. In the 1860s, an estimated four hundred fishermen lived in the boomtown of Caviar, now called Bayside. The town, dedicated to the catching and processing of sturgeon meat and roe, was viable enough to support a general store, a post office, and a rail station. The station transported some of the harvest to New York City; the balance was delivered by steamship to markets in Philadelphia.

Overharvesting, coupled with a growing water pollution problem, ended the industry. Despite catch restrictions initiated in 1904, the industry dwindled, and all commercial harvest ended shortly after World War II.

Today, nearly one hundred years after the population showed signs of stress, sturgeons remain uncommon denizens of the bay, although there are recent signs they may be making a slow recovery.

But the poster child of the fishing industry, the bivalve that made the economy of Cumberland County hum, was the oyster industry. Historically abundant, finding in Delaware Bay the right mix of nutrients and salinity, oysters triggered a boom in fishing and commercial boatbuilding that supported thousands and was the resource that underwrote the construction and occupations of most bayshore towns—Dorchester, Newport, Cedarville, Fairton, Port Elizabeth, and, of course, Mauricetown.

You can mark the onset of this mollusk mania by looking at the dates stamped on the plaques proudly affixed to the houses in our town. Most were built in the early 1860s. While oysters were recognized as an important resource as early as 1719, when laws were enacted to protect the beds from overharvest, the real boom in 'ersterin' did not begin until the late 1800s, and it coincided with the Victorian period.

I don't know why this salty bivalve was the rage among Victorian men and women, but it was. An average of ninety oyster-laden railcars per week were leaving Port Norris–Bivalve, fortuitously close to the population centers of New York and Philadelphia, by the 1880s. More than three hundred schooners and three thousand men were involved in Delaware Bay 'ersterin', and the shucking houses and canneries operated with sweatshop efficiency.

The oyster industry was immensely profitable, and by World War I it had evolved into a $10-million-a-year enterprise. As noted earlier, at one time the bustling town of Port Norris boasted of having the highest density of millionaires in the United States, but the share of profits that trickled down to the average "shucker" in the factories was truly a study in Upton Sinclair economics. Wages were meager, and many of these factory workers were African American. If you visit Heislerville, Port Elizabeth, Dividing Creek, Cedarville, and Mauricetown, you'll find an overwhelming preponderance of Caucasians. But Port Norris still contains a large number of African Americans whose fathers and grandfathers worked in the shucking houses of nearby Bivalve.

The vessel that was the utilitarian link between the mollusk and its Victorian consumers was the bay schooner, a sleek, shallow-draft, two-masted ship whose design harks back to Europe but whose lengthened hull, sweeping shear lines, and heart-shaped stern bear the refining stamp of Delaware Bay.

The shipping industry, which saw the construction of more than five hundred vessels between 1870 and 1935, made Cumberland County the second largest shipbuilding county in the state (after Camden County). It, and the coastal trade it spawned and supported, propelled the captains of these vessels to social and economic prominence. To look at the large, opulent Victorian structures that flank the banks of the Maurice River in our town is to peer through a window onto the region's boom past.

The first time I saw these stately homes, in the town I would one day call home, I saw the ships, too. A dozen or more. Resting at anchor. Tugging at their lines as if they wanted, more than anything, to be off once again with the tide.

They were unmasted, of course, converted in the early part of the century to motorized propulsion. But they had not lost any of the sleekness in their lines, and they were all that remained of the once mighty oyster fleet. What the distance did not disclose was the rot that infected their hulls, the same rot that had gripped the region after the industry collapsed.

In the 1950s a parasite called MSX attacked the oyster beds and ruined the fishery. The region's economy went with it. Today, a state-funded laboratory works to develop virus-resistant mollusks, and what success they have supports a vestigial oyster industry. But the days when more than 300 white-sailed ships plied the waters of the bay are over. Today, the grandsons of these oystermen make do with what the water offers and consumers are willing to buy.

HAULIN' NET

At 10:15, Tom drew up just off Bay Point and killed the engine. People who throw dice with the universe are foolish to put all their money on a single cast, so in the spirit of "doing it all," Tom and John finish their day by hauling net set for menhaden, known locally as bunkers.

They aren't a food fish, but they are legal, and their current numbers make them commercially viable, barely. Schooling fish, running on average ten to twenty inches long, they are used primarily as bait—for sport fishermen hoping to tie into a striped bass and for recreational crabbers (the ones who are not using chicken necks in their crab traps).

And bait for Tom and John. The menhaden he uses to bait his crab traps Tom nets himself, helping to keep his overhead down. He sells the surplus mostly to local sport shops.

Tom and John took up positions in the stern and started to haul, drawing the net out of the dark water just as fishermen have done for centuries. It was hard work but pleasant work—with both men smoking and trading quips. In the net, piling up on the deck of the boat, were sleek, silvery forms of fish. It took about five minutes to haul the 180-foot net, and then the more delicate work began. Tom and John stood to either side, slipping the net into a waiting blue barrel with one hand and slipping fish through the threaded openings with the other two.

I know humans have only two hands, but I don't see how they could do what they were doing with any fewer than three (and I was watching).

Unlike the crab traps, which are baited, gill nets constitute a passive capture system. The net is anchored at one end and swings free in the tide. A swimming fish reaches the net, tries to push through, and, meeting resistance, tries to draw back. The gills, the protective flaps that cover a fish's ventilation system, then become ensnared and the fish is held—unable to go forward, unable to retreat. The fish are mostly torpid, the black pupils in their white eyes giving them expressions of incomprehension—a fish in the headlights look.

The bunkers are transferred to waiting peach baskets. The first haul netted (literally) two bushel baskets of fish. Tom scowled. "I dunno what happened," he said. "I tol' my wife this morning we weren't goin' to do no good today."

Clearly two bushels wasn't a sizable catch. Very probably bunkers weren't going to "pull them out," as Tom had hoped.

The other net was not far away, and once again Tom and John went through the centuries-old exercise. This effort proved slightly more successful than the first, largely because the first net was 180 feet; this one was 225.

In the end, the catch amounted to five bushels of menhaden. There was a time, and not long ago, when Tom would sell thirty bushels of menhaden at the local bait shops. His recent average was ten.

I did some quick math. Calculated the bunkers at the rate of $10 a bushel, added $130 for the crabs. Measured that against Tom's $250 a day cost estimate. Came up red. Tom, apparently, had come to the same conclusion.

"Well, that was a total disaster all day," he announced, and he wasn't grinning. Moon's beatin' us today," he said. "We never do well on a full moon. I dunno why."

While John washed down the boat, Tom lit up another cigarette and retreated into the wheelhouse for the short trip home. I followed.

"So what would you do if you couldn't turn a profit fishing anymore?" I asked.

"I dunno," he said. "Figure somethin' out," he added, grinning. "Gettin' borderline, now," he amended.

"What do you figure the problem is?" I said, leaving the question open. Wondering how Tom would frame the challenge he and other watermen were facing.

"I dunno," he said again. "Six billion people," he said, taking the broad view, surprising me. Then he got more specific, more tribal.

"All I know is that there's enough for all of us," and there was an exaggerated volume in his voice that toed the line between stubbornness and conviction. Once again I was brought to wonder how much Tom knew or guessed about my background. "I just don't want any more change."

There was a pause, then Tom spoke again, summarizing all his waterman's pride and frustration.

"We don't want anything given to us, but we're tired of things being taken from us. And I don't remember any time when they took somethin' and gave it back.

"That's it in a nutshell," he said, saying all he or anyone working the water needed to say.

We were back at the dock by 11:10. It took another ten minutes to hose down the boat, stow the gear, get the day's catch into the back of Tom's pickup, and go on our separate ways. John to go bass fishing. Tom to the various outlets to sell his catch and get home in time to work on his son's boat.

Me? I went home to transcribe notes and think about the difference between people who engage the natural world with their hands, like Tom, and those who engage it on paper, like me.

Both of us united in our love for the bayshore; both of us counted among the 6 billion people on planet Earth. The 6 billion people who are all competing for the planet's resources. Like Tom, I believe that there is indeed "enough for all of us."

But what happens when the world's population hits 7, then 8 billion?

Like Tom said, it was "borderline, now." And even if Delaware Bay is a long way from famine, and even if the people who live here are as resourceful as Tom, it still remains that we, all of us, are adrift on the same planet. Just like the nutshell that is Delaware Bay, all the resources we have are the ones that are left.

CHAPTER 4
Seventh-Month Itch

The Fourth of July had come and gone. Unlike those other summer holidays—Memorial Day, whose celebration was shifted for convenience, and Labor Day Monday, whose timing was plotted for convenience—the day the delegates of the American colonies, meeting in Philadelphia in 1776, proclaimed freedom from tyranny is still celebrated on the Glorious Fourth.

But since the Fourth fell on a Friday this year, everyone got another three-day weekend anyway.

New Jersey played a key role in the American Revolution—serving as a battleground, a staging area for the Continental army, and a source of men and supplies. The battle for Trenton is widely regarded as a turning point in the war insofar as Washington's rout of Hessian troops on the day after Christmas 1776 revitalized his forces and supporters. A defeat at Trenton would very probably have led to the collapse of America's bid for independence (and the Fourth of July might be known as Insurrection Day instead of Independence Day).

The trials that Washington and his troops suffered during the eight-year war for independence are part of our cultural inheritance, and if asked about this, most Americans think automatically of Washington's winter at Valley Forge, Pennsylvania. Few Americans (few who are not New Jersey residents, anyway) realize that the commander and his troops spent not one but two equally challenging winters bivouacked just outside Morristown, New Jersey.

South Jersey's role in the Revolution was more supportive than direct, although the region did not lack for sons of liberty. In fact, in Greenwich, on or about February 22, 1774, a band of local patriots engaged in a "tea party" in league with, but less heralded than, that famous tax revolt in Boston Harbor.

As the story goes, the British brig *Greyhound* landed a quantity of tea in the port of Greenwich, where it was transferred, secretly, to the basement of the home of a crown loyalist. Empowered by the "articles of association" as established by the Continental Congress, a committee convened on Thursday, December 22, in the patriotic hotbed of Bridgeton, where a subcommittee of thirty-five individuals was appointed to deal with the matter.

Too late. The next day the committee learned that a subsubcommittee of Greenwich patriots, acting somewhat independently, had already exercised their patriotic prerogative and, disguised as Indians, burned the tea. The Greenwich Tea Party is commemorated by a monument erected in 1908.

Growing up near Morristown, I knew all about New Jersey's pivotal role in the American Revolution (although I confess I didn't learn about the Greenwich Tea Party until many years later). Nevertheless, the Fourth of July was a holiday I greeted with severely mixed feelings.

Don't get me wrong, I loved the pageantry and patriotism. Loved the hokey, homespun township parade, the local fair and the fireworks that followed—institutions I'm sad to say survived no longer than my youth. But the Fourth also marked an ugly truth that all the bunting could not hide. The Fourth was a temporal milestone: the quantifiable, unimpeachable proof that a part of my glorious summer recess was already over.

Two weeks earlier, when the last school bus of the year dropped me at the end of our street and my feet raced each other home, summer had seemed eternal. A whole, splendid, and unblemished span of days during which I could get up when I wanted, explore wherever I chose—catch turtles, find birds' nests, play inningless baseball until it was too dark to see a pop fly and fireflies would emerge from the shadows to test the quickness of our hands. I loathed school, so in balancing measure, loved the freedom of summer.

The Fourth of July was the first chink in summer's armor. Its arrival signaled that some measure of those precious summer days had already passed. If some of it could pass, then all of it could pass. That meant . . . that would mean . . .

Summer would end! A new school year would begin.

The horror of this realization tied my stomach into knots that no number of hot dogs, hamburgers, or Popsicles could undo. It has now been nearly forty years since my feet last sprang from the top step of a school bus. Even so, even now, it is a rare Fourth of July when I don't experience a moment of sick anxiety and absorb the body blow of understanding.

That summer is not eternal. That freedom does not last.

MAKIN' HAY

The gate was open, but I parked this side of the posts anyway. First, because it's a bad policy to park on the other side of locking gates that you do not have a key to. Second, because while I had walking privileges on Bill Garrison's salt hay farm, I'd never specifically asked for, or been granted, the right to bring a car onto the property. I didn't want to presume too much.

Even before stepping from the car, I reached, reflexively, for the bug shirt that lives on the back seat May through August (when it's not in use). Those parts that are not ventilating mesh are 100 percent tight-weave cotton. With hood up and screen face mask zipped closed, it's as close to in-sect-proof as a garment can hope to be.

I picked it up and then, with reluctance, put it back. In-stead, I donned a baseball cap, my sole protection.

"It's pretty cool today," I reasoned. "And a week later in the season. Be fewer bugs, maybe."

But the real reason I left the bug shirt behind was to save face. Fact is, sweatin' and swattin' flies is just part of hayin' on the bayshore. And hayin' is a tradition that goes back as far as the first Europeans.

THE DIKING OF NEW AMSTERDAM

After the Dutch disposed of the Swedes in the upper reaches of Delaware Bay, they inherited a landscape that triggered their sense of enterprise and begged for their en-gineering skills. Low and inundated by the tides, the marshes flanking both sides of Delaware Bay constituted thousands of flat, arable acres—arable as soon as they could be ditched, diked, and drained anyway.

Coming from a low-lying country that they had wrested from the sea, the industrious Dutch saw the extensive marshes as farmland waiting to happen. They set to work turning wetlands to upland, and when the British took con-trol of the region in 1664, they not only were content to let the Dutch continue their fine reclamation efforts but also adopted, modified, and expanded those efforts until, ulti-mately, much of the bayshore—from Salem to Cape May County—was highly productive pasture and farmland.

Far more productive than the sandy and nutritionally

impoverished soil of the pine barrens. Mercifully free from the plow-busting rocks that the retreating ice sheet had left as a parting gift to would-be farmers in southern New York and Northern New Jersey, and to future suburbanites, too.

My grudge against the last ice sheet of the earth's most recent glacial period is personal. Two decisions separated by ten thousand years sentenced me to weeks of servile labor and engendered in me a fine hatred of geology. The first was the universe's determination to end the southern advance of the Laurentide Ice Sheet on the shore of glacial Lake Passaic, a large body of water that once covered much of what would one day be called Morris County, New Jersey.

When the universe sounded retreat, the ice sheet, rather than carting back the accumulated debris topped off from several old mountain ranges, elected to leave it behind—depositing an amalgam of pulverized stone and glacial-turned rocks that geologists call a "glacial moraine" and that future residents would call Long Island and Northern New Jersey.

The second decision was made by my father, Gerald W. Dunne, who in 1956 purchased a newly built ranch house at 57 Washington Avenue, the precise location that ten thousand years earlier the ice sheet had dumped what I judge to be its biggest load of rocks. My early years were spent trying to undo the lazy mischief inflicted by a mile-high wall of ice so my father could have the lush, green, flat suburban lawn that all postwar suburbanites dreamed of. Knowing nothing about glaciers, or their capaciousness, I came to believe that it was the curse of all young boys to exhume and haul an endless parade of rock-filled wheelbarrows while Saturday morning cartoons went unwatched.

Ultimately, my parents surrendered to the futility of the effort and brought in loads of overlying topsoil (which of course had to be spread)—but not before instilling in their oldest son an abiding hatred of rocks and anything to do with them.

If anyone wants to know why I am not a geologist, and do not own a wheelbarrow today, he need look no further than my parents' backyard.

Ironically, the same glacier that undermined my childhood enriched both my adult life and the bayshore. After the Delaware River punched a hole in the Kittatinny Ridge, forming the Delaware Water Gap, it carried in its swirling waters the fine, glacier-ground sediment that settled upon and added rich acreage to the shores of Delaware Bay.

Not all at once, and not without a great deal of give and take. The deposited soil and the plants that took root in it struggled to keep their heads above water. They managed this by compacting the roots and stalks of vegetation into peat, whose depth increased as generations of plants flourished and died. The ocean and bay countered this defensive maneuver with rising water levels as all that water locked up in the receding ice sheet melted and flowed into the world's seas.

The product of this give-and-take is tidal marsh, a substrate that is for part of the day dry land, part of it wet. It takes a very special kind of plant to survive, much less flourish, in a world as ecologically polarized as tidal marsh, and it's not surprising that the plant that stepped into this ecological breach was a grass—one specialized to take root in anaerobic soil, draw its nutrients from seawater, and at the same time defy the moisture-sucking osmosis of seawater. The name of this supergrass is *Spartina.*

Over the course of thousands of years, the marshes evolved in the no man's land that lies between open bay and dry upland. As they grew, they changed in both profile and vegetative cover—the lower lying, outer marshes that are typically inundated twice a day by the tide; the older, higher, and drier upper marshes that are reached only by the highest moon- or storm-driven tides. The classic grass of the outer marsh is *Spartina alterniflora,* or cord grass.

It has no commercial value, but it does have immense value to wildlife. If you have ever driven on any of the causeways leading out to the towns and beaches of the Jersey Shore, this, by and large, is the verdant, green ocean of grass that you pass through.

Higher up, flush against the upland forests, is a different kind of meadow, characterized by a different brand of grass. This is *Spartina patens.* Not so tall, tightly packed and fine, it looks like hay; in fact, it is hay—salt hay. And when those early colonists arrived in the New World, they, and their cattle, discovered that it made a wonderful, nutritious fodder. Better still, not only did domestic animals prefer salt hay to traditional hay but salt hay was almost maintenance free. Ordinary hay, over time, leaches the nutrients out of the soil, requiring applications of fertilizer. Salt hay takes its nutrients right from seawater.

As the Dutch surmised, and as growing numbers of English bayshore residents discovered, if a person were to dike and ditch lower marsh and control the amount and duration of its immersion, the low marsh, with little commercial value, could be transformed into salt hay–producing high marsh, and animals could graze on it.

Salt hay has commercial applications beyond mere grazing. Harvested and stored, it can be used for stable bedding and feed in winter. Ropes can be made from it. Roofs thatched. Icehouses insulated.

In the nineteenth and twentieth centuries, when Cumberland County's glass industry flourished, salt hay was used to pack glassware. During World War II, salt hay became an important component in the concrete used in the construction of airport runways. Today, owing to its inability to germinate in any but tidal habitat, it is used as mulch in gardens and newly seeded lawns and roadsides.

If you were to drive through the bayshore communities today, you would find a dearth of livestock and thatch-roofed barns to house them. The glass industry has been greatly downsized and cheaper, synthetic packing material found. Refrigeration has eliminated the need to store winter ice.

In the late nineteenth and early twentieth centuries,

there were more than twenty families commercially harvesting salt hay in New Jersey tidal wetlands. Today, there are . . .

"The Darrien brothers in Sea Breeze, Mick Coombs in Fairton, Dean Berry in Port Norris, and Marshal Hand in Goshen," said my unimpeachable source, a levelheaded, slow-talking resident of Dividing Creek.

"Didn't Marshal die last year?" I asked.

"Yeah, he did," my source agreed, with a thoughtfulness that had less to do with faulty recollection and more to do with reluctant resignation. "But his brother's workin' the farm."

Counting the Hands, that made four. Counting my source, Bill Garrison, it made five. Five families still plying a trade nearly as old as the European presence on the bayshore and now nearly endemic to the region. It was Bill Garrison's property I was on, and his farming operation I was going to see.

Bill Garrison Jr., son of Bill Garrison Sr., is a Dividing Creek native—a boast that is not uncommon among "Crickers." It was his father who bought the 550 acres of upland "neck" and surrounding marshes that constitute most of the family farm today. This acreage, acquired in the early 1980s, augmented the 50 acres already owned by his grandfather. All three generations of Garrisons cut hay.

"My grandfather remembered when cuttin' was done with horses," Bill recollects. "But by the time my dad was hayin', it was all done with tractors."

And still is. Bill, age forty-four, his brother, Fred, and

now son, Matt, soon to be sixteen, run their mechanized operation from the last week in June until November. Starting at 7:00 A.M. and working until late afternoon, Monday through Friday, the fine salt hay is first cut and raked into windrows. After five days of drying, and providing weather and tides cooperate, the hay is baled, loaded onto flatbed wagons, tarped, and stored. Ultimately it will be trucked and sold to garden centers in North Jersey, New York, and Connecticut, where salt hay is still prized as mulch.

"How much hay do you harvest a year?" I asked one day in early July, the day I'd arranged to watch the operation.

"Fifteen thousand bales," Bill said matter-of-factly. Despite the weight of his disclosure, it lifted the corners of his mouth.

"Lot of work," I couldn't help noting.

"We plan to do it as long as we can," he replied.

The reason we, meaning he, had time to talk during hay season is that the baler had run out of baling wire and haying was stymied until his brother, Fred, could get back with another spool.

"It's always sumthin'," Bill said, philosophically.

But the delay gave me time to ask my questions, make arrangements to return, and spend a little time testing my reflexes against the defensive skills of the evil greenhead flies that are the bane of all things that bleed on the Delaware Bay marshes.

In the old days, they used to cover the undersides of the horses pulling the combines and wagons with tar-soaked burlap bags and place burlap bags on the horses' heads. Despite these precautions, the horses' legs would run with blood. Humans suffered, too, but I noted that while Bill and Fred and Matt wore jeans, their upper parts were guarded only by T-shirts and baseball caps.

"Bugs don't seem too bad this year," I said, to certify my hardiness.

"Last week was pretty bad," Bill's voice answered from somewhere beneath the baler.

I shivered, as a horse shivers when a greenhead bores a hole in his flank, and wondered just what kind of insect onslaught would get a "pretty bad" rating from a third-generation Cricker.

I would probably call it "intolerable." You would probably call it "life-threatening."

After the baler was back on line, I took some pictures, then took my leave, motivated more by the heat that was building up in my long-sleeved shirt than by the attack of insects.

"You goin' to maybe load some hay with us when you come back?" Bill chided gently.

"Sure," I said, wondering whether I meant it.

MATT'S JOYSTICK

It's about a mile out to the open marsh where the hay is stacked and the equipment parked when it's not in use, and it was a beautiful day. Sunny and clear with a pleasant, fly-dampening northwest wind coming in off Turkey Point. I don't know what Bill Garrison Sr. paid for the 550-acre tract, but whatever it was people probably laughed at the figure at the time. Twenty-five years later, in an age when New Jersey is approaching the "build out" point, developers are probably crying.

It's a beautiful piece, much of it forested upland, home to red and gray foxes, otter, great horned and barn owls, and a healthy herd of deer, kept healthy by hunting.

In the 1990s, much of the region's coastal wetlands were bought up by the power company—part of a deal struck with the state. It was determined that the cooling towers for the nuclear plant near Salem were killing weakfish fry. It was decided that, rather than spend hundreds of millions of dollars to build new cooling towers, the company could "mitigate" its damage by "restoring" wetlands to tidal flow. Mitigation, to a large degree, meant gaining title to marsh diked for salt hay, knocking down the dikes, and flooding the marsh.

The result was that lots of man-made high marsh was transformed into man-made low marsh, and many of the animals that thrived in high-marsh habitat (like breeding and wintering northern harriers, like breeding saltmarsh sparrows, like wintering short-eared owls) lost out.

Bill Garrison Sr. refused to sell. If short-eared owls and saltmarsh sparrows choose a man of the century, you'll find Bill's name on the short list.

About ten minutes after I arrived, a Ford pickup broke from the trees and pulled to a stop. Fred was at the helm, Matt in the passenger seat.

"We'll be balin' over where you were the other day," Fred explained, as he made his way toward the New Holland baler. "You're welcome to ride with Matt or walk."

"I'll ride," I said, walking the short distance to the flatbed hay wagon hitched to the smaller tractor.

"Thanks for letting me come along," I said to Matt over the throaty chug of the tractor, which sprang to life just about the time he vaulted into the seat. Already as tall as his father, Matt had gentle, intelligent eyes and moved with a sure, confident grace.

"I'll help load if you like," I heard myself say.

"I don't need any help," he said, with the defensiveness of a fifteen-year-old doing a man's work who seems suddenly challenged by a stranger.

My mistake.

"I mean I'd like to help, if you'll let me," I corrected. "Show me what to do and I'm game," I said, wondering if it was true.

"Okay," he answered quickly, and not because he was fifteen and I was an adult but because he'd evaluated the offer and found it acceptable.

I realized only later that the reason I was so impressed by this display of easy confidence was that I've grown so used to a world in which even adults, and much more teens, are beset with diffidence.

Matt eased the tractor into gear, and we lurched forward, off the road and onto the flattened track that led out to the marsh. His driving was as good as his judgment. Maybe better. That remained to be seen.

The wind blew the tractor's exhaust back into my face, and my first instinct was to shift to the other corner of the wagon. What I did instead was inhale deeply and smile.

Fact is, this wasn't the first time I'd sat on a hay wagon and been bumped and tossed over fields. But it'd been a long time. I was a couple of years younger than Matt. My brother Dave and I spent several weeks on an Ohio farm owned by a World War II buddy of my father's and pen pal of mine. His name was Earl Marlatt.

In a life now filled with memories, those weeks on the Marlatt farm still rank among the greatest. From that experience, I learned more about life, and its connectedness to nature, than I did with all the schooling that came after. The curriculum included being awakened before dawn, eating first breakfast, doing chores, eating second breakfast, workin', eating lunch, doing more work, eating dinner, going to bed, getting up, and starting in on whatever it was that didn't get done the day before.

We never argued to stay up. There wasn't any TV, and in just a few short hours, someone would be shaking us awake in the dark.

Breakfast was eggs, fatback, homemade bread with honeycomb, steak, pancakes, pie, and anything left over from dinner.

Lunch, the big meal, included several kinds of meat, potatoes, fresh bread, pie, and anything left over from breakfast.

Dinner was anything not eaten at lunch, and no matter how much we ate (and we ate plenty), there were always leftovers.

I never ate so much in my life. But I never worked so hard in my life.

We cleaned stalls. Fed chickens, cattle, hogs. Repaired fences. Barked posts. And best of all, helped cut, bale, stack, and store hay.

I learned to drive a tractor—a big John Deere you started by spinning the flywheel. I was handed a .22 rifle, with a 4× scope, and given the sacred duty of keeping the gladioluses safe from marauding groundhogs. I fired my first shotgun—a ten-gauge double with three-and-a-half-inch shells that should have put me right on my keister (but to everyone's surprise, including my own, it didn't).

But I think the most valuable thing I learned was that hard work that really means something isn't exhausting, it's empowering. And that if you work past the point when your brain says you can—by tossing hay bales until your arms are numb and your body is on fire from the needle-fine shards of grass boring into every pore—there isn't a person in the world whose eye you cannot meet. Because there is not a person living who has endured more than he can.

This makes you equals.

Earl raised cattle and gladioluses. And while he remained a bachelor until the day he died, he played a big role in raising two boys.

TOTE THAT BALE

We reached the field after a five-minute drive. Fred was already baling. The sunlight-colored blocks were lined up at twenty-foot intervals, waiting to be placed on the wagon. This is where the mechanized age of farming ends and 4000 B.C. begins.

Matt drove past the first bale and stopped at the second. Dismounting from the tractor, he retrieved the bypassed bale. Slung it, lengthwise, and set it down at the front of the wagon, just right of dead center. He snugged the next two up against it, and I saw the sense to it before he explained.

"We'll do five across; three loading on this side, then two."

"Got it," I said.

Without another word, like a man who'd explained this a million times, he slipped out of his work gloves. Handed them to me. Moved back to the tractor.

"My God," I whispered through the smile I found impossible to keep off my face, "I'm fifty-six, and I'm loading hay on a wagon again."

Don't throw your back out, I reminded myself as I reached for the first bale. Not only would doing so be a betrayal of Matt's trust but it would be sure to earn me a wifely admonishment when I got home.

A lot has changed in forty-three years. Standard transmissions are no longer standard (in fact, not even an option on most cars). Americans who fly on commercial jets no longer laugh at those silly Russians who were the butts of our jokes because they were always standing in line. When people say "China," nobody means "Taiwan."

One thing has not changed, and that is a bale of hay. It is just as heavy and cumbersome as it always was. Just as unforgiving if your aim is poor and you need to adjust midthrow. Just as itchy.

And it smells just as good. Sort of puppy-breath fresh and bread-coming-out-of-the-oven wholesome. It surprised me, I think, that salt hay, when all was said and done, smelled and felt and was just like normal hay. It tastes a little saltier (of course I tried it!). Maybe that's why cattle that are not on low-sodium diets prefer salt hay over alfalfa three to one.

I didn't throw my back out. I set those bales side by side, three deep, end to end. When one side was filled, Matt directed the tractor to pass on the other side of the waiting bales so I could easily fill the waiting slots. The first layer was down almost before I broke a sweat.

Matt stopped the tractor. Stepped to the wagon. Flung two bales up, center one lined up like the ones before, second one set at a right angle to both the first and the ones below. I saw the sense in this, too. Capping the first layer crosswise would help anchor the load.

"Want me to take over?" he offered.

"Nah," I said, grandly. "I'll finish this one."

The bales were no heavier, but the lift was one bale higher. I was sweating by the time the second layer was done and happy to call it quits when Matt announced, "Need five more on top. Anyplace."

Anyplace was just about how they fell.

Feeling more proud of myself than I deserved, I watched as Matt pulled the wagon out of sight and wondered whether I had the gumption to load another.

We'll never know. I did start the second. But halfway through, my efforts were interrupted by the arrival of "the Boss" as Matt shouted, nodding.

A grinning Bill Garrison strode up. Shook hands. Took over.

By that time I was plenty ready to turn things over to the professionals and watch from the sidelines the easy banter and fluid maneuvering of father and son as they made quick work of the remaining bales on the field. The rest of the afternoon would be spent shifting bales into the trailers for transport. It would be dirty, itchy, backbreaking work, and I was more than glad that I had writing to do.

But before we headed off to our separate tasks, while Bill went to coordinate afternoon plans with Fred, I had a chance to ask Matt a bit about himself. Learn the kinds of things a writer might want to include in a book intended, in part, to honor dying traditions because readers, already estranged from the land, may wish to know them.

I found out that despite his obvious athleticism he wasn't interested in organized sports (or computer games) but that fishing was what he did on the two days a week when he wasn't working for his father.

I learned that he liked rock music, especially the Red Hot Chili Peppers and Foo Fighters, and that he didn't have a girlfriend (yet), but he did already own a truck, a blue-green 1996 Ford F-150. He was, on this beautiful day in the summer of the year and the spring of his life, just days away from getting his farm (driver's) license. A privilege (or, really, entitlement) that suburban sixteen-year-olds envy without understanding, or recognizing, the basis for.

But the thing I most wanted to know was precisely the thing I was most hesitant to ask. I was afraid because I wanted the answer one way and feared it would go the other. Once again, there was something about Matt's confidence that supported mine. I got the question out, and, better than this, I got the answer I wanted, too.

"So, Matt," I said, remembering our earlier miscommunication, and choosing my words with care, "what do you expect to be doing ten years from now?"

"Doin' this," he said, quickly. "With my dad," he added confidently.

Cutting and baling and stacking and wholesaling a product that has been part of the bayshore's heritage for nearly four hundred years. Maintaining and repairing a dike system that is in constant need of work against rising sea levels and marauding muskrats. Defending the precious acres of salt hay from encroachment by *Phragmites* grass and periodic buyout offers. Facing an ever-dwindling market and ever-rising costs.

Confronting a world that more and more seems less and less like theirs.

Off the cuff, because I'd run out of questions and because I could see Matt's father coming, I asked, "What do you want people to think when they think about you?"

I thought I was getting used to the directness of Matt's answers, but I was surprised again.

"That he's honest," he said.

What a great kid, I thought as I started back for my car, and Matt and Fred and Bill started the hard and itchy work that would command the afternoon.

Matt Garrison. Age almost sixteen. Cricker and salt hay farmer. If he never leaves Dividing Creek, he's destined to go far. In fact, he's already got a great head start. Four hundred years and three generations.

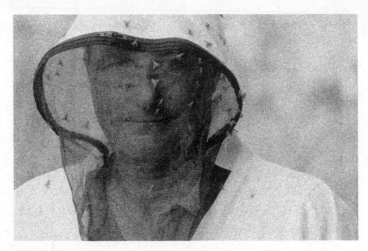

CHAPTER 5
Kudos for Our Air Defense System

RED ALERT

I parked in the usual morning spot on Turkey Point Road, leaving two wheels on the pavement. Max and Raven, who are intimate with the steps in our ritual morning walk, got to their feet and immediately started pummeling the inside of the car with their tails. In the language of Labrador retrievers, this means "Walk time, now!"

Between September and mid-May, the next steps are reflexive: I step from the car, crack the hatch, and unleash two hundred aggregate pounds of pent-up canine energy. But in the summer months, there is an added step in the morning ritual; a precautionary habit, painfully learned, that saves time, frustration, and blood.

I sit in the driver's seat and listen for the telltale pings of biting flies ricocheting off the car.

If the pings are less than incessant (meaning you can hear individual pings), the volume of insects is tolerable—providing you're dressed for it. But if the rattle of insects throwing themselves against the car has the consistency of an audible blur (meaning there are so many pings that it sounds like a cross between a German M-42 machine gun and a hailstorm), then the morning walk at Turkey Point is canceled. No questions asked.

This was one of those mornings. Hot, humid, and buggy beyond the imagining of any but a bayshore resident. Even with a hooded bug shirt, I wasn't brave enough to face it or cruel enough to subject my canine duo to it.

"Sorry, dogs," I said, "change of plans." We drove the extra ten minutes to Bevan Wildlife Management Area. A forested, inland tract that lies far enough from the marsh to be mercifully bug-free—although one incautious step off the packed-sand road and into the vegetation would result in a world-class harvest of ticks.

No. That's not a contradiction. Bugs, loosely but technically speaking, are insects. Ticks and chiggers are arachnids in the order Acari, in families respectively called hard-backed ticks and harvest mites. What these phylogenetically distinct groups have in common is their choice of hosts.

This is us.

The Worst Biting Insects on the Planet

As projects go, the writing of this chapter is one that I have eagerly anticipated. No discussion of a Cumberland summer can be complete without a thorough exploration of the hordes of biting insects that pretty nearly run things on the bayshore during warmer months. If you want to know why the bayshore is so unpopulated, you must, of course, calculate in the agricultural sterility of the pine barrens to the north and the buffering advantage of Delaware Bay to the south. You also need to consider the population-siphoning impact of nearby Philadelphia and, of course, the economic doldrums that beset the region.

But unless you figure in the population-dampening impact of biting insects, you cannot understand the bayshore, its history, or its people.

How few there are, how hardy they are. Unless your hide is made of Kevlar or your pores excrete 99 percent deet, I'd think long and hard if I were you about choosing to live here: the place I cheerfully and confidently dub "home to the worst biting insects on the planet."

Fair Sampling

I know what you're thinking, you residents of Florida and Texas and Alaska. You're thinking of the clouds of mosquitoes that infest your states, and you're right—to a point. You *should* be perversely proud of your biting-insect hosts. I've sampled the worst you have to offer.

Snake Bight Trail in the Everglades, where mosquitoes are as thick as smoke. Aransas National Wildlife Refuge, where I've worn a bandanna over my mouth in order to breathe. Alaska's North Slope in July, when caribou are driven to migrate hundreds of miles to escape the blood-sucking hordes.

And you're right! You've got world-class mosquitoes. I bought the post cards; I've given blood.

But mosquitoes, about twenty-five species, are the least of the insect challenges faced by residents of the bayshore. We laugh at mosquitoes here. Linda and I frequently go camping on the North Slope of Alaska in July rather than face the plague of insects we'd have to face at home.

There is only one place on the planet where I have been driven to run, really run, from biting insects. It was a dirt road just south of Greenwich, New Jersey. I had on a long-sleeved cotton shirt but had neglected to wear a hat. On the Delaware Bayshore, in late May and June, when the straw-berry flies come out, this is suicidal.

What's a strawberry fly?

Man, you don't know nuthin' about biting insects if you don't know what a strawberry fly is. Make mosquitoes seem as benign as mayflies.

THE WONDERFUL CLASS OF INSECTA

Actually, and in perfect truth, most of the hundreds of thou-sands of insects on the planet are as benign as mayflies or dragonflies—benign to humans, anyway. Dragonflies, in their larval stage, are ferocious predators of small pond creatures and, as adults, catch and consume smaller insects.

And many of the creatures that fall into the phylogenetic

construct we call "the insects" are greatly loved by humans. They include butterflies, moths, honeybees, and crickets (to name just a few).

In quick and simple review, all insects are part of the very large scientific phylum Arthropoda, which is then broken down into subphyla. This is where flying bugs, like mosquitoes and strawberry flies, diverge from other biting critters, like ticks and chiggers. Mosquitoes and biting flies are in the subphylum Mandibulata—a large and diverse grouping that includes all the insects. That's right, the mosquitoes you hate are anatomically similar to the butterflies you love. Both groups are sucking insects.

Chiggers and ticks belong to the subphylum Chelicerata—a less diverse but hardly limited group that includes species as outwardly dissimilar as horseshoe crabs, spiders, and scorpions.

It is interesting that both the Mandibulata and the Chelicerata are distinguished by—in fact, named for—their differing mouthparts. Not only is this important from the phylogenetic standpoint but it also goes a long way toward explaining how differently these creatures, while united in their vendetta against our species, inflict their torment.

Chiggers and ticks latch on. Mosquitoes and biting flies bore a hole or snip the skin, but they don't latch on. From a behavioral standpoint, chiggers and ticks and mosquitoes and biting flies also differ in their mode of attack. Mosquitoes and flies, as flying insects, seek you out. Chiggers and ticks, which do not fly, wait in ambush. You find them.

But bushwhack and clamp on or home in and bore, the consequences are pretty much the same. Your lot in life has been reduced to the status of a meal.

How Can Something So Small Inflict So Much Torment?

In Mauricetown, as everywhere in temperate regions governed by the change of seasons, people look forward to the days when temperatures are mild enough to lure us outside. Linda and I have a backyard to call our own. We have a porch with Adirondack chairs so perfectly suited to a sedentary existence that, had they been designed before our ancestors climbed down from the trees, our species might never have known an upright existence.

On those wonderful, warm evenings of May, we like to sit in those chairs and watch the sky change colors over glasses of wine. But we're Cumberland County residents. We never let our guards down completely. We monitor the wind, and as soon as it begins to falter, our Insecta antennas go up, because sure as God made little bitty things, the no-see-ums are going to rise.

"Got one," one of us will say, not even bothering to explain what "one" refers to.

"Yep, got a few here, too," the other will reply.

Then, without another word, drinks in hand, and accompanied by the scrape of wood on wood, we retire to the screened-in, no-see-um-proof porch.

With alacrity.

The only sure defense against these shock troops of the

biting-insect season is to put yourself behind a screen barrier whose mesh is so tight it darn near inhibits the passage of oxygen molecules—and a biting insect that is almost as microscopic.

No-see-ums are minute, biting midges or flies; members of the family Ceratopogonidae and closely related to black flies—a group of biting insects at home where Adirondack chairs are made.

How minute? Some species are as small as a millimeter long. A trophy-class no-see-um is about three millimeters, about as long as the comma found in this sentence. But the measure of irritation they dole out is worth an exclamation point at least! Falling short of pain (it's nothing to say "ow" over), the bite is nevertheless sharply irritating, and what the tiny harpies lack in force of chomp they more than make up for in numbers.

No-see-ums arrive in invisible clouds. They are so tiny that even a modest breeze can hold them at bay. But as soon as the wind drops, they appear like evil magic and coat exposed body parts in a patina of exquisite itching torment that increases until your tolerance evaporates and you bolt indoors.

They don't stop at the cloth line either. No-see-ums are wonderfully adept at crawling down collars and up untucked shirts. They seem particularly adept at winding their way through hair and feasting on the scalp below.

People with curly hair swear that they are prejudiciously targeted by no-see-ums, but I tend to doubt this. Only females bite, requiring, as most biting insects do, the protein offered by a blood meal to produce eggs.

The eggs, larvae, and adults (when not foraging) are found in wrack, or tide-borne vegetation matted on the higher parts of the salt marsh. Adults are active day and night. They seem not to travel more than several hundred yards from the marsh that supports them. They have two population peaks, one in late April and May, and another, smaller emergence in late August and September.

They can ruin a family picnic and stop a volleyball game at point serve. They can stop a romantic moment faster than you can say, and almost as fast as you can execute, coitus interruptus.

My worst experience with no-see-ums occurred at Reeds Beach—a place famous for its concentrations of red knots in the spring. I was participating as one of a host of volunteers at a scheduled "shoot," whose objective was to net and band numbers of red knots and ruddy turnstones. The net launching was, unfortunately, very successful. It resulted in a couple of hundred birds pinned beneath the net.

I say "unfortunate" because, when the net veiled the birds, the wind dropped to nothing. In mid-May, at the peak of their emergence, the no-see-ums swarmed. Inflicting biting, itching, maddening torment upon the hair, ears, eyes, noses, and exposed parts of biologists and volunteers.

It was like performing a delicate operation in a hair shirt infested with acid-injecting lice. I saw people injure birds in their torment and frenzy. I never participated in another shorebird shoot again.

The good thing about no-see-ums is that they die easily (anything approaching a swat is overkill), and for most people the resulting itch is short-lived. They can be defeated with a layer of baby oil (in which they get mired), and they are, to some degree, discouraged by deet-based insect repellent.

But I've doused myself with deet and still been driven to near madness by no-see-ums. Fortunately, no-see-um season is brief on the bayshore, and the wind is our ally. Compared with what comes next, no-see-um season is not even an essay in the art of torment.

STRAWBERRY SEASON. YUCK!

If you Google "strawberry fly," you'll find scant reference to this Cheerio-size harpy—but one of the first will be to the website of a Cricker named Bernie Sayers, whose ten-point test for Downe Township residency begins with this challenge:

"You are probably from Downe Township if you . . .

1. Know the difference between a greenhead and a strawberry fly. Bonus if you can tell which arrives earliest in the season."

"Season," of course, refers to biting-fly season. The answer to Bernie's bonus challenge, should you not know, is strawberry fly—a biting fly that emerges and reaches plague proportions in late May through June (just about the time strawberries appear in farm stands). Greenhead flies emerge later.

Strawberry flies are among the Tabanidae—a family of biting, clear-winged flies that includes horseflies, greenhead flies, and deer flies (which strawberry flies closely resemble). Slightly smaller, slightly plumper, slightly darker bodied, they pack the same painful punch as their more widespread brethren, using the same pain-inflicting instruments: a pair of scissorlike mandibles that they employ to snip the skin of their hosts in hopeful anticipation of a blood meal to follow. The bite is painful enough to draw your hand quickly, and the animals are soft-bodied enough that a semiforceful swat is all it takes to dispatch one. That's the good news. The bad news is that, during strawberry fly season, even da Vinci's Vitruvian Man would be about a dozen hands short.

If it's a bad year, if you are on a wooded neck bordering marsh and it is between late May and late July, you are going to discover what every Cricker (and bayshore resident) knows, which is:

You are standing in the wrong spot. Strawberry flies are invariably found in numbers falling between a swarm and a plague. They seem programmed to concentrate on their victims from the waist up (with particular attention paid to the area about the head, face, and neck).

Not particularly fast to bite or avoid a swatting hand, they are nevertheless overwhelming in their tenacity and the sheer weight of their numbers. They feed only in daylight, seem attracted to motion, and are not commonly found far beyond the ecotone that separates the forested uplands and open marsh.

I'd never heard of strawberry flies until I moved to South Jersey, and I did not really become acquainted with the insects until 1979, the year I conducted my first season-long survey for breeding harriers on the bayshore. What I received was a postdoc in biting-fly research.

The methodology, as outlined in my proposal, was easy enough. Patrol the edges of tidal wetlands by car. Stop at strategic points. Get out. Scan for harriers. Follow them with binoculars until they led me to their nests. It was a good strategy. Flawed only because the study period was late May through early July (strawberry fly season), conducted on the Delaware Bayshore (the universe's strawberry fly stronghold), and I would be standing outside.

I'm a very stubborn person and have a high threshold for pain. Both these qualities were tested that year. Finding harriers on open marshes requires lots of scanning, and holding the image of a hunting male harrier in binoculars until he captures prey and carries it to the female waiting at the nest may take an hour or more. That's an hour with your hands and face exposed.

I learned, over the course of many torturous hours, how strawberry flies seem particularly drawn to the soft skin lying at the junctions of the fingers. I found, after a day's survey, that my hands were so swollen I could not close my fingers around the steering wheel of my car. I drove home palming the steering wheel and gearshift.

Since I grew up with deer flies, the bites of strawberry flies don't leave a lasting welt or impression on me. In less than ten hours the swelling went down and the torment could start anew—which it did. For the several-week length of the survey.

Strawberry fly season lasts just about as long as strawberries are found on farm stands (which is also about the time it takes harriers to fledge their young). Their numbers diminish just in time for the niche to be filled by deer flies—whose bite is just as hard, but whose numbers, while impressive, never seem to reach a level that prompts a person to run. Deer flies just try to outlast you. In places like Turkey Point, Hansey Creek, Backneck, and Greenwich, you can swat them into September.

THE MOST FEARED PREDATOR ON THE MARSH

The last time I wore shorts was July 1978. That was also the year I found a Wilson's phalarope on the dikes of Brigantine National Wildlife Refuge and started dating Cynthia. Yes, the phalarope and the fashion statement and the relationship are all related.

A bird flew in front of my car, a white 1978 VW Rabbit. It disappeared into the marsh. My companion, an attractive young birder named Cynthia, and I sprang from the car and trained our binoculars on the spot where the bird had landed. Both of us were wearing cutoffs. What the two of us endured over the course of the next twenty minutes cemented our friendship.

Another way of saying we became blood kin. The green-head flies that infest the salt marshes of New Jersey from July into September cut us to pieces.

By the time the bird reappeared, our legs were running with rivulets of blood. The hood of my car, upon which we'd climbed for the protective elevation it afforded, carried the indented evidence of our evasive efforts until I traded it in.

I learned several things that day. First, Cynthia was one tough birder, a keeper. Second, the greenhead hunting pattern targets elevations between ankles and belts (perhaps so as not to directly compete with strawberry flies). Gain a little elevation, and you gain a defensive edge. Third, only an idiot wears shorts in summer in coastal New Jersey, an idiot or someone who has never been on the receiving end of *Tabanus nigrovittatus*.

Like strawberry flies, greenheads are armed with scissorlike mandibles that they use to inflict a painful cut. When strawberry flies bite, you say "ow." When greenheads bite, you use expletives.

Greenheads are larger than strawberry flies—ranging from one half to slightly more than one inch long. Shaped like bullets, capable of flying thirty miles or more, living several weeks as adults, they seem less inclined to hunt in packs but evidence more tenacity and more cunning than your average biting fly.

I know that entomologists will cringe at this assertion. As evidence, I offer the reaction of my fly-hating dog, Raven, whose whole life has been a study in making the planet more suitable for Raven.

Strawberry flies, Raven dislikes but tolerates. Greenheads, she fears and loathes. Put one greenhead in the same enclosed porch as Raven, and the ensuing life-and-death struggle is epic, with the greenhead maneuvering to land on

Raven's unprotected posterior regions and Raven whirling like a dervish and lunging and snapping with her jaws whenever the evil thing flies near.

When a greenhead hatch is on, female greenheads emerge from their three-month period of development to find a blood meal, mate, and deposit between one hundred and two hundred eggs in the mud or soft earth surrounding the salt marsh. Laughing gulls gather by the hundreds over hatches, snapping up emerging greenheads by the thousands.

I've seen adult laughing gulls return to their nests and regurgitate fist-size lumps of fresh-caught greenheads. Raven's not that skilled. But in a one-on-one, she gives more than she gets.

Greenheads are attracted to motion and dark colors—qualities that make black Labs, walking on open salt marsh roads, prime targets. I'm also suspicious that greenheads are heat-sensitive. I've noticed that, when my dogs are swimming, greenheads seem particularly attracted to their projecting heads—seem suddenly to be more numerous and more ferocious than when the dogs were standing on dry, hot land.

Whether they're cunning or no, heat-seeking or not, nobody is going to argue against the greenheads' celebrated toughness. While no-see-ums can be snuffed with a brush and strawberry flies give up the ghost at a swat, greenheads seem armor-plated and gifted with more lives than a cat.

When you swat a greenhead, you learn to plant your hand and roll the insect beneath applied pressure. Even if you hear the crunch or snap that in the case of a strawberry fly means "game over," you learn never to assume a greenhead is dead. I've seen flattened greenheads shrug off a mighty swat and fly on. I make it a point, whenever a greenhead rolls from my hand and onto the ground, to plant a foot and give it a twist, just to make sure.

Too often efforts to dispatch greenheads fail to reach this happy point. In addition to being tougher, greenheads are also quicker than strawberry flies. It gets worse. As air temperatures go up, the insects' reflexive response time goes down. Some bayshore residents claim that greenheads can read minds, assessing precisely when a raised hand is about to fall (and fly out of harm's way).

I won't subscribe to this, but I won't gainsay it either. One thing I know: sometimes the only way to nail a particularly quick and tenacious greenhead is to sacrifice. Allow the bastard to bite you, get his greedy little scissors planted in your hide, and slow his response time with the taste of blood.

Sometimes it's enough. And the satisfaction that comes of killing a fly that has wounded you is almost enough to justify the bite. But it's not enough to get me to wear shorts on the bayshore in July.

LOATHSOME

Paul Kosten, father, school psychologist, bayshore resident, and friend, grasped the offending arachnid between index finger and thumb and removed it with a deft and practiced pull. "See this?" he said, brandishing the female deer tick under his son Dylan's nose. "This is the only creature on the planet that it's okay to feel good about killing."

I don't necessarily agree with this sentiment. I've killed more than my share of greenheads with considerably more satisfaction than remorse. But very few humans this side of an extremely committed Buddhist would take Paul to task for his condemnation of these accomplished, fascinating, bloodsucking relatives of spiders and scorpions.

Ticks, whose bites can, and frequently do, transmit serious and even deadly diseases to humans and other animals, and whose complex cocktail of digestive juices gives rise to an exquisite and enduring itch, will find very few defenders among those of us who fall into the ranks of hosts.

Like mosquitoes and biting flies, female ticks need a blood meal in order to produce eggs. Unlike biting insects, ticks attach themselves to their hosts. They need time to feed. They get this by being stealth feeders.

Ticks don't announce themselves by buzzing around your head. They wait in places where you are likely to pass, latching on when you brush against the grass or leaves that hold them. They can also sense you standing below, by either reading your heat signature or detecting your carbon dioxide exhaust. A silent drop onto your head is all the air time a tick ever gets or needs.

Ticks don't commonly start their excavations in places where they are easily detected (or reached). Eight-legged, mobile, and fast, ticks crawl quickly and purposefully toward some special place—onto your head under a shielding mat of hair, or behind your knee, or into your crotch.

Compounding the detection problem, depending upon

species and stage of development, ticks can also be small to minute. The larvae of the black-legged tick (commonly called the deer tick) is the size of the period at the end of this sentence (but less visible because it is pale, not black). A nymph, the second stage of the tick's development, is the size of a poppy seed. An adult female deer tick is no bigger than a sesame seed.

Also unlike those of biting insects, the bite of a tick is painless—the barbed harpoons that the animal implants in your skin go unnoticed. Commonly, the first indication a person has that a tick is embedded in his hide is the resulting itch. At this point, the animal might still be attached or it might have dropped off to lay its eggs and finish its life's work. You get credited with an assist.

Cumberland County's all-star tick lineup includes the dog tick, the lone star tick, and, of course, the deer tick. Dog ticks are usually active only in warmer months. Deer ticks are active as soon as the temperature approaches fifty degrees Fahrenheit. In Cumberland County, New Jersey, this means any month of the year (and I've been bitten by ticks every month of the year).

How common are ticks on the bayshore?

Linda's and my record is one hundred ticks per leg, collected while navigating one hundred feet of a leaf-strewn trail in midsummer—a total of four hundred ticks. This averages to a population density of four ticks per linear foot.

Since that day, we've never stepped off paved or unvegetated sand roads in summer in Cumberland County. Ever. Biologists of my acquaintance who must go into the woods in summer almost invariably impregnate their clothing with Permanone before a submergence.

What's Permanone? It's nothing you want to serve your houseguests. In fact, the instructions on the chemical killing agent direct users to apply it only to their clothing and to avoid contact with skin.

There are, of course, precautions that can help deter ticks. Among them: wear white clothing so ticks can be seen and removed; wear long-sleeved shirts and pants; tuck pant legs into socks; check yourself thoroughly after being in a tick-infested area.

Embedded ticks can be removed with tweezers. Small ticks can be pried off with duct tape.

Me? I just stay out of the woods.

But one thing I will absolutely, positively not do, *ever*, is walk through tall grass, on the bayshore, from May through September. Tall grass houses torment. The price of cavalier trolling is at least one week of abject misery.

It means . . .

CHIGGERS!

I have only once threatened a fellow human with death (and meant it). It was a physician, a person driven by oath and conviction to alleviate human suffering. I made my threat after the healer, when informed that I was suffering from an acute infestation of chiggers, wrote a script for an over-the-counter topical ointment.

"No," I said, unbuckling my belt. Exposing legs so covered with itchy red bumps that even he must have itched.

"You will give me industrial-strength cortisone or I will kill you, now."

He did, and his well-conceived action saved two lives.

I got chiggers in Cumberland County, along the railroad tracks off Route 555. I recall walking hurriedly through a single, small patch of grass no larger than a tetherball court. Twenty foolish seconds; one week's misery.

A chigger is a near-microscopic red mite. Some people claim to be able to see them; I've only seen their handiwork. Magnified, they look like your worst science fiction movie nightmare. Undetected and untreated, they will make you wish you were never born.

I hate chiggers. Hate 'em. Hate 'em. Hate 'em.

They serve no function this planet cannot happily do without. They undermine the whole concept of a benign God.

Pestilence, death, taxes I can accept. Chiggers constitute a malediction too far.

If the Christian Creator had infested Job with chiggers instead of boils, He would have very quickly appreciated the limits of human frailty and patience. As a species, we have certainly been guilty of many affronts to nature and the divine order, and a reckoning in the here or the hereafter is due.

There is still no excuse for chiggers.

Like ticks, chiggers must latch on to their hosts in order to feed. Unlike ticks, and the other pests already described, chiggers do not suck blood. Chiggers are meat eaters.

Drawing upon all the objectivity I can muster, let me say

that their feeding behavior is pretty amazing. Being small and less toothy than ticks, the animals must seek out some soft, wrinkled place to work, often attaching themselves to the base of a hair follicle. Once affixed, they spit up a tiny amount of enzyme that melts our skin, which the creatures then suck up through a straw whose hardened walls are formed by the enzyme and our molten hide.

They do not burrow beneath the skin. They do remain for two to three days, and, unlike ticks (whose flat, armored bodies seem impervious to pressure), chiggers are killed or dislodged at the first scrape of a fingernail across the bite site.

It's already too late. The enzyme and foreign protein are already implanted. Our own immune systems do the rest.

With me, and most people, it takes twenty-four hours before chiggers make their presence known. Twenty-four more before the full extent of the infestation becomes evident. Twenty-four more before the full brunt of the misery is upon us. Then, from day four to day seven, the itching misery slowly diminishes. By day ten, we may actually feel like we can lead a normal life again.

Chiggers have the particularly nasty habit of homing in on tight, skin-to-skin or fabric-to-skin locations. This means ankles (if you are a sock wearer). This means around your waist (where underwear elastic rests). For unliberated women, this also means bras.

There have been times when I could identify the bra Linda was wearing the day she was anointed just by connecting the itchy red dots.

It's small compensation, but sufferers should know, at least, that humans are not the targets of chiggers. We're victims of mistaken identity. For chiggers to be effective (that is, to finish their meals unmolested), they need hosts whose immune systems are not going to trigger an alarm and a violent reaction.

There are species of chiggers not native to North America that do use humans for hosts. Their bite is unobtrusive. But in Cumberland County, New Jersey, you have two choices. Stay out of the grass and woodlands, or douse yourself with insecticides or sulphur powder and take a scrubbing shower within a few hours of a possible exposure. Do not, under any circumstances, wear the same clothes without washing them. And do not, under any circumstances, come near me.

Share a bed with a person exposed to chiggers, and you'll share the joy.

DISCLAIMER

There are, certainly, places in the world whose insects constitute a greater health threat than Cumberland County, New Jersey. Insects whose bites are more painful, insects whose bites transmit diseases that are crippling or fatal.

Not my argument; not my point. I still assert that the Delaware Bayshore is home to the worst plague of biting insects on the planet—our own population-dampening Air Defense System.

But I hasten to admit that the bayshore's red zone is not only just skin deep, it is geographically restricted. It extends hardly any distance inland. Go two or even three miles away from coastal areas and your torment will be considerably reduced. When a Millville real-estate agent or a representative of the county department of tourism tells you, "Oh, there are no bugs here," so long as he or she is not referring to coastal areas, you can believe it.

And if you do choose to move to the bayshore, you can always do what Linda and I do. Come bug season, we get out of Dodge. Take a vacation to the White Mountains of New Hampshire (where you have to contend only with black flies) or Alaska (where surviving is only a matter of dealing with mosquitoes).

Wait a minute, you're thinking. He didn't say anything about mosquitoes.

I told you. Here, we laugh at mosquitoes. But in case you are interested, there are plenty. And, unlike Alaska, we have them all year.

CHAPTER 6
Party Time!

It was only a hundred yards from Higbee's Marina Café to the docks. Bolstered by a three-stack of pancakes, bacon, and coffee (and carrying nothing but a daypack and writing pad), I elected to walk. The guys burdened with multiple rods, tackle, and supersize coolers crammed with iced-down beverages, of course, unloaded their vehicles at the dock.

"Goin' with us today?" a shirtless and harried young man asked in the consonant-eating dialect of the bayshore.

"Yes," I said.

"Need a fishing rod?" he asked of the slab of fish he was cutting into bait-size slivers. A perfectly understandable, and no doubt perfunctory, question given the nature of his job, which was mate on a party boat. Coupled with my obvious objective, going out on said fishing boat. Supported by the obvious absence of anything resembling fishing equipment in my hands. My answer surprised him enough to stop the fillet knife in his hand midstroke.

"No," I said. "Just going as a passenger, but I'll pay full fare. How much is it?"

"Forty dollars," he said, recovering quickly.

"How long we going out?"

"Six hours," he replied, dropping his knife, hurrying to assist several guys who were struggling to get a near-refrigerator-size cooler off the dock and onto the fifty-five-foot head boat *Miss Fortescue*—one of a half dozen boats taking paying customers out to fish the waters of Delaware Bay.

Ten years ago, there'd been considerably more boats in the fleet. And during the summer season, roughly late April through November, they were all running at or near full. That was back in the days when Delaware Bay was crammed with seatrout, or weakfish ("weakies" as they are known locally) and the village of Fortescue boasted of being the "Weakfish Capital of North America." In those halcyon days, a party-boat fisherman could hardly drop a jig off the side of a boat and not hit a weakfish on the head.

"We'd drop lines, and everyone would have a fish on," Jim Higbee, the *Miss Fortescue*'s captain said, flatly, accurately, and wistfully. "Then we'd get ahead of the school and hit 'em again." This was back in the seventies and eighties, when Jim was just starting out as a party-boat captain and when the *Miss Fortescue*'s hull was laid. Running party boats is all Jim Higbee, now approaching fifty, has ever done, all he's ever wanted to do, a conviction he inherited from his father and grandfather, who were party-boat captains before him.

Running experienced and novice fishermen alike out onto the waters of Delaware Bay. Getting them over fish. Giving people of all ages a day of fun and excitement on the water. It's just a lot more fun and a lot more exciting when you catch fish. Of the thirty-five people who came aboard that day, I was the only one who didn't have that ambition.

At least not this time on the water.

A LITTLE BIT OF RAIL TO CALL YOUR OWN

I made my way to the stern, taking a position where the great human drama called "going fishing" would unfold. There are lots of ways people engage the natural world, but moment for relaxing moment, and pound for edible pound, there are few activities that can vie with fishing.

If you read the promotional material sponsored by the sport-fishing lobby, you'll find market-fanned estimates hovering around 50 million fishing enthusiasts in the United States. Ranked North America's third most popular outdoor activity (behind gardening and bird watching), fishing takes almost as many forms as there are fish species, and it is an avocation that attracts people from all walks of life, without regard to sex, age, race, religion, income level, IQ, or belt size.

You can be the president of a Fortune 500 company, take the company jet to Montana, spend a week at an exclusive lodge, cast a fifteen-hundred-dollar fly rod over the waters of the Ruby River by day and talk about it over a single-malt Scotch that evening. At the core of this exercise is the ambition to catch fish.

You can be a retired Mississippi sharecropper, walk down to the banks of the Pascagoula River, and dangle a shrimp-baited hook from a cane-pole rod for perch while sipping encouragement from a paper bag–wrapped bottle. Your objective, too, is to catch fish.

Former President Jimmy Carter is an avid fisherman. He has, in fact, written books on the subject. Ernest Hemingway was a fisherman. He, in fact, won the Nobel Prize for Literature with a "fishing story" first published in September 1952 in *Life* magazine. The story was entitled "The Old Man and the Sea."

Chuck Yeager is a fisherman, Eric Clapton is a fisherman, Christ's apostles were fishermen, and I was (and sometimes still am) a fisherman.

Curiously enough, about the only person in the world who was not a fisherman was my father, Gerald W. Dunne. Yet one of my greatest memories of him is the time he took me fishing.

And how all-inclusive is that! Fishing is so universal that even people who are not fishermen are fishermen!

Most of the people clambering aboard the *Miss Fortescue* fell somewhere between the Orvis fly rod and cane-pole classes of fishermen and somewhere between Ernest Hemingway and my father in terms of an avocational commitment. Nevertheless, and as far-reaching as these parameters are, a person didn't have to be very discriminating to note that the people onboard were hardly representative of the North American, much less fishing, community as a whole.

Most were male. Most were in their late forties to mid-sixties. Almost all were with a family member or group of friends. Almost none wore long pants.

Shorts and T-shirts were the uniform of choice. Baseball caps and running shoes were near universal. Most had brought their own rods and tackle. A dozen or so rented rods from the boat for five dollars. Bait (referred to as squid but in reality cut-up shark or bluefish) was included in the forty-dollar fee.

Typing people, in this age of this world, is a risky undertaking, but I'm going to do it anyway, and I'm even going to do it in a somewhat unorthodox and almost un-American fashion. I'm going to look at my shipmates from a Marxist standpoint, using, as Karl Marx did, their relationship to the "tools of interaction" as my determining point of reference. That tool is the *Miss Fortescue.*

The Socialist philosopher Marx, as you may or may not know, concluded that there were two classes of people on the planet: the workers (or the proletariat) and the entrepreneurs. What distinguished these two social classes was their relationship to the "tools of interaction." One class of people, the entrepreneurs, owned the tools. The other class, the workers, used the tools.

By this measure, the people buying passage on the *Miss Fortescue* were members of the working class. They didn't own the boat. They were just using the boat. They were "working-class" fishermen.

Plenty of people who go saltwater fishing do own their own boats—and most who do will agree with the definition of a boat that even Karl Marx would subscribe to: "boat, n. A hole in the ocean, surrounded by fiberglass or wood, into which one pours all his money."

While some party-boat fishermen fish in this manner because they are economically savvy, many, and perhaps most, buy a place at the rail because party-boat fishing they can afford, and owning a fishing boat they cannot.

Aggregation is one of the elements that puts the party in party-boat fishing. Economy is what puts it in reach of almost everyone, and by party-boat standards, *Miss Fortescue* is a bargain.

As one money-conscious fisherman, a gentleman who was built like a chest of drawers and who supports his fishing habit by working as a clerk, confided to me, "It costs sixty dollars to go out of Atlantic City or Cape May but only forty dollars here. An' here you're fishing half an hour out of the dock. It takes an hour out and back before you're fishin' at those other places. I get more fishing for my money here."

Note, he said "fishin'" not "fish." It's a small, but important, distinction.

Only three of the thirty-five patrons onboard were women. This didn't particularly surprise me. There was only one young person onboard, a tousle-haired, first-time fisherman of fourteen. This did surprise me. Why? Because it was, after all, a weekend—prime time for family bonding experiences. Because summer is all about having fun, and having fun is what fishing is all about.

And because when you ask a person on a party boat how he started fishing, he will tell you, almost without exception, that his father or grandfather took him. There seemed to be at first, and second, glance, too, an unbalanced number of mentors versus apprentices onboard this beautiful Saturday morning.

At precisely 7:00 the captain fired up the engines and the *Miss Fortescue* eased out of her slip. Five minutes later, we cleared the channel and headed south-southwest into the bay.

"Aren't you down' any fishin'?" a guy wearing wraparound sunglasses and a Philadelphia Phillies cap (turned backward) asked.

"Probably wet a line later," I lied.

"Oh. Want a beer?" he asked, and without waiting for an answer reached into his cooler for a Sam Adams.

And that's the other thing that puts the party in party boat.

Just about the time the pancakes and bacon in my stomach were coming to uneasy terms with the unexpected and near-unprecedented appearance of twelve ounces of carbonated malt beverage, the captain reached the northern edge of the shipping channel and cut engines, and thirty-five party-boat fishermen sent thirty-two ten-ounce weighted lines to the bottom of Delaware Bay.

The three in noncompliance were me; the fourteen-year-old, whose bait-casting reel had spun a world-class tangle of line known universally as a "crow's nest"; and his father, who was trying to help restore order.

The delay hardly mattered. After ten fishless minutes, the captain ordered "Pull 'em up," and we went off in search of more productive waters. Found them, too.

"Fish on," a voice exclaimed. "Fish on; fish on," several other voices chorused, and the announcement had the quality of an electric current as it moved around the rail, a visceral pulse that triggered both excitement and envy.

As the lucky angler worked his prize up from the waters, and neighboring fishermen craned their necks to see what he'd hooked, the mate, long-handled net in hand, moved into position.

"Fluke!" a voice intoned, and once again the electric pulse lapped the ship. Fluke, or flounder, was the target fish of the day and a fish on triggers a delicious dichotomy.

While a fish caught by someone else means one less fish for you, it also means that there are, de facto, fish below. Where there is one fish, there are usually more. As the fish was boated, and deftly unhooked by the mate, everyone else doubled and redoubled their focus on the lines entering the dark waters—the lines that would signal the gift of fish with a tremor or a tug.

Any residual envy was soon replaced with a mixture of sympathy and relief.

"Oh, tell me it's eighteen inches," the angler pleaded, as the mate adjusted the length of the fish on the measuring board. Curious-looking creatures, specialized for a life of foraging on the sandy bottom, flukes are elliptically shaped and pancake flat. Brownish olive above, white below, they possess a most curious adaptation, the lateral migration of one eye, along the dorsal side. Born as fish that swim on a vertical plane (as most fish do), flounder, as they mature, incline morphologically and behaviorally toward life as bottom-feeding fish.

This means they list, then fall on one side.

Rather than spend its time looking at nothing but the sandy bottom, the eye on the increasingly blind side migrates dorsally until both eyes are on the top side of the fish, situated just behind the mouth. It's a funny adaptation, but it works.

"Seventeen and a half" was the honest and unwanted verdict of the mate.

"Awww," the man who found himself one half inch short of a keeper fish moaned.

"But that's with its mouth closed," another fisherman, with an acute sense of fish anatomy and a loose interpretation of state fishing regulations, offered.

"I'll bet you can get another half inch if you press on it," yet another suggested.

But the New Jersey regulations are strict, and captains of party boats who want to avoid fines adhere to the letter of the law. The fish was a half inch short of the limit. It was returned to the bay.

A majority of the fish caught fall into this category. But less than half a mile from where the *Miss Fortescue* was drifting, the regulations are even more restrictive. In Delaware waters, the limit for fluke is 18½ inches; in Mary-

The ringing cry of the laughing gull is the quintessential sound of summer on the Delaware Bayshore. Here, an adult in breeding plumage adds his (or her) voice to the chorus.

ABOVE: *Bird watchers and ornithologists call it the clapper rail. But on the bayshore, everyone knows the bird as the "mud hen."*

BELOW: *Small but energy rich, the eggs of the horseshoe crab are the tiny loaves that feed the multitudes of northbound shorebirds.*

RIGHT: *With an entire food-rich tidal ecosystem to forage in, even the noisome willet cannot resist the nutritious bounty bound up in a tiny horseshoe crab egg. While migrating shorebirds depart by early June, the crabs continue to breed into July, providing food for breeding birds such as willets.*

BELOW: *In twenty years, the* rufa *subspecies of red knots went from a population high of 160,000 to fewer than 18,000. The commercial harvest of horseshoe crabs, whose eggs were consumed by knots to fuel the last stage of their migration, led to the decline. Shown here, one of those surviving 18,000 knots (along with a dunlin).*

ABOVE: *Dunlins in breeding plumage on the beaches of Delaware Bay. By late May, these Arctic nesters are in the final stages of their countdown before liftoff. Next stop, tundra.*

OPPOSITE TOP: *Haulin' (crab) traps on Delaware Bay. A waterman's life is hard and too often frustrating, but as the smile on Captain Tom Pew's face attests, he loves his work. Mate John Burens, shaking out the contents of a crab trap, is more enigmatic but no less dedicated to his craft.*

OPPOSITE BOTTOM: *Sport fishermen head out for a day on the water. Sure looks like it's going to be a hot one!*

ABOVE: *It looks like hay and it is. Salt hay! Jersey-fresh, sun-cured, baled, and awaiting shipment to a garden center near you (providing you live within a half day's drive of Dividing Creek).*

BELOW: *Young Matt Garrison, fourth-generation "Cricker," tops off a wagonload of salt hay on the family farm.*

ABOVE: *Everyone deserves a favorite place on the planet. Here's mine: Turkey Point, End of the World, New Jersey.*

RIGHT: *Hungry young in the nest appear to have this adult black-crowned night-heron working overtime.*

TOP LEFT: *If not for my trusty bug shirt, you can be certain that these strawberry flies would wipe that smile off my face. Welcome to the home of the worst biting insects on the planet.*

BOTTOM LEFT: *The evil greenhead fly, the most feared predator on the marsh. Wear long pants or be prepared to pay a price in blood.*

BELOW AND OPPOSITE: *Two sunrises on the Maurice River. Great sunrises are among the uncelebrated benefits of being a bird watcher.*

Two broods down and a long migration ahead. But in August, even adult barn swallows get to sit and take a break.

Birds abound on New Jersey's Delaware Bayshore, including this great egret. At the beginning of the twentieth century, egrets were extirpated in New Jersey. After a century of protection, they and other herons and egrets are once again common.

Maligned by some biologists, phragmites, or plume grass, not only provides nesting habitat for least bitterns, the very rare nigrescens *subspecies of swamp sparrows, and marsh wrens but also supports large roosting concentrations of migrating and wintering birds. It is also visually appealing (as Linda's photo attests).*

ABOVE: *Harvesting tomatoes is hot, hard work. Pictured here, one of New Jersey's first-string pickers, showing, literally, the fruits of her labor as well as her partiality to New York baseball.*

OPPOSITE: *In case you've ever wondered why New Jersey is called the "Garden State," here's a truckload of underscoring explanation en route to Lillian's Market.*

ABOVE: *The nimble, acrobatic, and vocal marsh wren is almost never out of earshot on the marshes of Delaware Bay.*

OPPOSITE TOP: *Their populations once severely depressed by DDT, ospreys are again common on the bayshore. The birds perched on this artificial nesting platform attest to the success of restoration efforts as well as the fishing skill of their parents.*

OPPOSITE BOTTOM: *This juvenile osprey (showing pale-tipped feathers) is watching the sky for fish. Of course, Mom or Dad will be ferrying it.*

Captain George Kumor of Heislerville, Delaware Bay waterman and conservationist, heads in with his day's catch. Note the trailing entourage of gulls: George's minions.

land, anglers can keep no fluke measuring less than 19 inches.

To add to the disappointed angler's disgruntlement, it wasn't long before his fishing buddy landed a fluke of his own—this one measuring out to 18½ inches.

"What kind of bait are they using?" an intense thirty-something angler on the other side of the boat asked.

"Gulp! shrimp" was the answer. Up and down the rail, anglers changed bait, bummed bait, or bemoaned the frugality that had prevented them from having the bait they wanted now.

"Gulp! shrimp," by the way, is not a typo. It's the name of this year's newest and latest "designer bait." An artificial bait, advertised to be impregnated with four hundred attractive scents, Gulp! shrimp is the newest achievement in humanity's never-ending quest to get the upper hand on fish. A quest that has pushed the average cost of catching fish up to around twenty dollars an inch on the basis of equipment alone (near as I can tell).

From rods so sensitive they can pick up a fish's vital signs, to lines as invisible as water, to lures that swim more enticingly than the bait fish they replicate, to fish-locating sonar that makes finding fish as easy as watching TV, to . . .

Equipment bags, tackle boxes, coolers, fish grippers, electric aerating bait tanks, nets, waders, vests, trolling motors, polarizing dark glasses, and . . .

"Gulp! shrimp," the guy at Higbee's Marina bait shop said, smiling. "This year's fad bait. Last year it was Gulp! squid. Year before that it was Gulp! shedders. Nothing really beats just putting a live minnow on your hook," he said with the kind of confidence that years behind the counter and an ample supply of minnows going for five dollars a pint confers on a man.

Gulp! shrimp, by comparison, were going for forty dollars a quart.

Of course, you don't need all of these fish-killing accouterments just to go out on a party boat. Forty dollars, plus five, gets you bait, basic equipment, transportation, and thirty years of fish-finding experience, courtesy of the captain.

But still, you do have to get to the dock—and given an average round trip of, say 120 miles in an SUV (real fishermen don't drive sedans), you're talking thirty dollars in gas at summer 2008 prices. Since you are getting up at o'dark thirty, you are going to buy breakfast on the road and probably purchase sandwiches, too. Figure another ten to fifteen dollars per person.

It doesn't include the ice you are putting in the cooler (two dollars a bag at the marina) or the soda and/or beer, or the sunscreen, or, should you choose, the price of having your fish cleaned (two dollars for flukes) or tipping the mate.

So even without shelling out forty dollars for a quart of this year's latest and greatest "fad bait," a party-boat fisherman can expect to pay a minimum of ninety to one hundred dollars for the privilege of maybe catching a keepable fish that will result, in the case of fluke, in four fillets.

Make note of the word *maybe*, because on the day we went out, there were eight legal flukes taken (two by one individual). This means that twenty-eight individuals aboard the *Miss Fortescue* went flukeless. Averaged out, using minimum figures, and an average size of nineteen inches per keeper fluke, the price of fluke on July 19, 2008, was about $22.00 an inch or, calculated at two pounds of fillet per fish, $191.00 per pound. The price of flounder, in the fish section of the supermarket most of the anglers would pass on their way home, was about $7.95 per pound.

Of course, if you factor in the mess of bluefish that were also boated, the aggregate cost per fish goes down considerably.

The fourteen-year-old first-timer, who, like most of those onboard, had yet to get a hit, dug into the stock of Gulp! shrimp. Hooked one through its artificial head. Lobbed it overboard. Overlooked the fact that the guys using this year's fad bait were backing up technology by putting a minnow on the hook, too.

If 400 attractive scents is good, 401 is better.

Then, like the thirty-four other like-minded individuals, the young fisherman settled into the timeworn tradition of fishermen everywhere: waiting for a bite. With nothing to do but nurse the intake of carbonated malt beverage, and be attentive to a tug on the line, minds are free to wander in time and space.

You can dream about the twenty-foot Boston Whaler you are going to buy someday. You can bemoan the frugality that led to your lack of Gulp! shrimp. You can replay old *Star Trek* episodes in your mind or reargue the argument you had about new porch furniture the night before.

Or you can do what I was doing. Casting the line of my thought back into the ocean of memory.

FISHING WITH DAD

The year was 1968, and the world—my world and the one run by the adults—was in turmoil. I was sixteen, a junior in high school, and if there is anyone reading these words who needs to be told that this is a very difficult thing to be, it can only mean that you are fifteen years old or younger or that you have selectively blocked all memories associated with being sixteen.

But on the morning that is anchored in my memory, nothing but good inhabited the world. I was sitting in the front seat of a vehicle whose heirs would someday be known as SUVs. I was sandwiched between the vehicle's owner, a gym coach, and my father, a fifth-grade teacher. We were on our way to Belmar, New Jersey, to join a party of sons and dads for a day's fishing on a party boat. The gym coach was the outing's organizer and, for some unaccountable reason, had invited my father and me along.

No matter how generous your standard, my father, Gerald W. Dunne, could hardly have been accused of being a fisherman. Likewise, he could deflect avocational labels such as golfer, bridge player, stamp collector, musician, square dancer, amateur astronomer—almost anything that might be construed as an interest or a hobby.

Curiously enough, he was good at sports. At the horse-shoe tournament that was the centerpiece of the Dunne clan's annual gathering, he and whatever family member was fortunate enough to be paired with him constituted the team to beat.

On the trip to Ohio that resulted in my summer on the Marlatt farm, Earl set up a water-filled coffee can on a hillside and marched us over to the far side of the valley. When nobody stepped forward to accept the challenge "Who wants to shoot first?" he handed the sleek, scope-mounted .218 rifle to my father.

Easing himself to a sitting position, anchoring his elbows just inside his knees, Dad lined up the can that was barely visible to the unaided eye. At the report of the gun, a geyser of water shot skyward. I could hardly believe my eyes.

"Ol' Dead Eye Dunne," Earl said, punctuating this pronouncement with a slap on his thigh. "That was his nickname in the outfit," he confided. "I'll bet he never told you that."

He hadn't. And it took me two shots to equal his performance—which is to say, I did not.

But despite his innate skills, and in defiance of the pastimes he pursued before meeting and marrying my mother, the man who was my father enjoyed nothing better than sitting on his patio in the summer and in his favorite living room chair in the winter. Sipping a beer from a small glass. Doing anything more physically engaging than the occasional game of catch in the front yard was alien to him and to our relationship.

Which made the morning we went fishing together even more delicious and memorable.

A few paragraphs back, I tacitly asked the question "Why would anyone want to spend so much money catching fish when it is so much cheaper to buy fish?" I'd like to address the question now, because it not only goes a long way toward explaining why it is that people go fishing but also is central to the memory of the time I went fishing with my father.

In essence, I think people go fishing not because it is fun, not because it is productive, not even because it is a hobby. I think people go fishing in order to escape, for a time, the life they lead and step into another. One that is, by the very nature of difference, more exciting than the one they left behind.

Fishing means getting up early. In the dark. At a time when other people sleep. There is something delicious about rising and knowing that the light in your room is the only one shining on the street. There is something bonding about speaking in whispers as you load the car and drive off into a night whose cloaking darkness is big enough for only two.

Do you remember the serene confidence that watching the dawn come up confers upon a person? Can you imagine (if you cannot recall) the empowerment that sitting down at a smoky counter filled with men imparts to a boy and how hearing the words "Order anything you like" vaults you into a higher plane of being?

One in which living on a tight budget and a teacher's salary has no bearing.

One in which a person can order bacon and sausage—both!—and not have to worry about a mother's (or a wife's) admonishment.

Then the smell of bait—oily, thick, and full of promise. The smell of the sea. The salty solution that is the primordial serum of our blood. The acrid smell of men who didn't bother to shower that morning so are not tainted by the neutering smell of soap.

The keening of gulls. The rumble of the ship's diesel engines underfoot and in your gut. The clank of metal rods in metal holders and the coarse talk of men whose tongues have been loosened by the absence, or near absence, of women.

The look. As they are suddenly conscious that you are among them, that they may have spoken too freely.

And the bonding assurance. Vetted by your father. Directed toward them but spoken to you. "He's old enough."

Suddenly old enough. What every young boy, deep in his heart, longs to be.

So fishing lets a kid step out of childhood just as it lets an adult step back into it. Away from the desk that pinned you against a wall of obligations all week. A boss whose name is found in the dictionary under "jerk." A relationship that was once a cup filled with promise and is now a vessel of routine. Money problems, health concerns, a car that's running funny, kids who seem to think that they're entitled to everything and that your sacrifices mean nothing.

But then you drop a line in the water. All existence draws down to two points and a single line. And at the end of that line lies possibility.

A fish to test your skill. A fish to ignite the envy of all around you. A fish big enough to fill the kitchen with stories when you get home that evening and the ears of coworkers and clients when you return to work on Monday.

It's not an escape from reality. It *is* reality. One so real that it can trump the unreality of the life you lead. Particularly if you are unhappy with that life you lead.

Which my father, apparently, was not. He had no reason to escape who or what he was, or if he did, he had an escape less obvious than going fishing.

We did catch fish the day he and I went fishing. In fact, we slaughtered 'em. Caught mackerel by the gunnysack. Took three, sometimes four at a time on diamond jig–weighed lines studded with "teasers." This was in the days before catch and release was fashionable. We took the fish home. Imposed them upon all of our neighbors and friends. Used the rest for fertilizer in the garden.

This, and the resulting stink, was probably the reason we never went fishing again.

I went over to the young fisherman slouched in the stern. A sandy-haired kid whose body was racing to catch up with his frame and whose pale eyes were furtive with unease. He and his father were part of a church group numbering about eight.

"Is this your first time on a party boat?" I asked.

"Yeah," he said to the deck.

"How do you like fishing?" I asked.

"I'd give it about a six," he said after a moment's thought.

"You know you're the sole representative of your age group onboard," I said.

He answered with a teenage-worthy shrug.

"What would you be doing today if you weren't out here fishing?" I said, guessing the answer in advance, steeling myself for it anyway.

"Playing games on my computer," he said.

I give him credit for being out here. Give his father credit, too. Out here together. Engaged in an avocation that has been, for many years, handed down father to son. In an age when the artificial reality that is cyberspace trumps the real-world experience that is fishing.

"Pull 'em up," the captain shouted over the loudspeaker.

Possibility, the captain figured, was somewhere else this morning.

THIS SIDE OF . . .
"Drop 'em," the captain shouted over the loudspeaker, cutting the engine. "We're right over them," he added, not really explaining what "them" was. Experienced fishermen concluded quickly and correctly that "them" meant bluefish (because finding fish on radar means running into fish massed enough to show an echo, and in Delaware Bay, in July, the only fish schooling *and* worth stopping a boat for are bluefish). Inexperienced fishermen simply dropped their lines as ordered.

The blitz was exciting but short-lived. The school of

nine- to ten-inch "snapper blues" moved out from under us, but a number of fish were boated, and ice-filled coolers were called into play. The fourteen-year-old boated two. The guy standing next to him, who had traded his flounder setup for a rod rigged up with a bucktail, caught six.

"You want them?" he asked the church group.

"Yeah, sure," they agreed.

Linda would kill me, I thought, if I brought home a mess of bluefish, and while this is something of an exaggeration, it is no exaggeration to say that she would have refused to eat them. Flukes are light, delicate, savory fish. Bluefish are vascular, and, even fresh, they have a strong taste (hence the bucktail fisherman's eagerness to divest himself of them).

Clearly, he and Linda enjoyed similar taste.

Weakfish, the onetime champion fish of Delaware Bay waters, is likewise lightly flavored and textured. For decades it dominated catches and drove the party-boat industry, bringing high levels of satisfaction to party-boat fishermen and a decent living for party-boat captains, like Jim Higbee.

A tall man with a beaver-pelt haircut and a passing resemblance to the actor Donald Sutherland, Jim Higbee moves with an easy, ambling grace and spends most of his time in the wheelhouse. Dealing with fishermen is the job of his deck hand.

"I used to have three or four hands back when business was good," he said.

"How long ago was that?"

"First bad year was 1990," he said. "Three years ago weakfish just collapsed."

"Why?" I asked, knowing it was the same question I'd posed to Tom Pew several weeks earlier, wanting a second opinion.

"Caught 'em all," he said, catching himself too late. It's not the kind of thing a party-boat captain is supposed to say to a writer. His quick follow-up observation is actually closer to the truth and in accord with Tom's diagnosis.

"I dunno," he said. "I jes' don't know."

Nobody does, truly, although everyone speculates. It could be any number, or combination, of factors. Like many fish, weakfish are cyclical. Boom and bust in fish populations is as natural as yin and yang.

But understanding the problem is not the same as remedying the problem, and even if a remedy is found, it is still not going to affect the fishing season in Delaware Bay any time soon. The drop in weakfish has seen the numbers of fishermen falter and the income of boat captains fall. The thirty-five fishermen on the boat today constituted the peak for the season. The boat holds forty-five, and back when the hull of the *Miss Fortescue* was laid, in the late 1970s, it ran full in summer—weekdays as well as weekends.

Nevertheless, Jim appears on the dock every morning. From April to November. Goes out when he's got enough customers to break even. It's all he's ever done, all he's ever wanted to do.

"Gotta start payin' attention to business," he said, shutting down conversation.

"Pull 'em up," he shouted into the microphone, and from the bridge you could hear the conjoined rattle of reels.

The wind had fallen, and it was getting hot. At the next stop, lines were dropped, T-shirts began to come off, and lunches and drinks appeared. Bags of chips were passed. Offers for brews or sodas made and accepted. Close proximity and shared interest constitute fertile ground for socializing and fishing, but the reputation party boats have as hotbeds of rowdiness and drunkenness is greatly exaggerated. Case in point: most of those onboard were drinking soft drinks. Most of the conversation would have been acceptable at a church social. I counted the number of expletives I heard. Not including *darns* or *shoots*, they numbered precisely eight.

Late-afternoon and all-night trips can be different. I recall a time I went out of Barnegat Light on a pelagic birding trip aboard the *Miss Barnegat Light*. The boat's sister ship, berthed next door, was going out for bluefish, and the conversation from the cadre of beer-enhanced gentlemen standing in the bow had reached politically incorrect levels. Every subject, in every sentence or sentence fragment, seemed bent upon fornication.

What the lively and unsuspecting louts had failed to note was that in our midst was a cadre of Amish women, who had in their care a number of young Amish birders. After a brief caucus, a posse of Amish temperance members stalked off our boat and stormed onto theirs. With hands on hips and fingers in faces, they read the beery boys in the bow the riot act.

I've never heard so many "yes, mahms," "no, mahms,"

"I'm sorry, mahms," and "We didn't know, mahms" before or since.

Around eleven o'clock the *Miss Fortescue* got over another school of snapper blues. One man boated a twenty-three-inch fluke (which held on to be the winner of the boat's pool of $155). The price-conscious gentleman who'd calculated that *Fortescue* gave him more fishing for the buck boated a nineteen-inch fish—which proved to be the first flounder of legal size he'd boated in two years of trying. He was thrilled, grinning ear to ear.

The kid?

I'll give him credit. He struggled valiantly, with both boredom and the lack of sleep. The problem with getting up at o'dark thirty is that it catches up with you. Even without beer, the rising heat, full stomachs, and attention-lulling absence of fish put those onboard into a torpid state. The rod, which had been raised, began to wilt. The stance, which once was erect, began to adopt a tired slouch.

I couldn't help but think back to my own first experience on a party boat. The fish I was pulling up three and four at a time. The captain saying loudly of what was certainly my luck (but what I presumed to be skill), "Now here's a young man who knows how to catch fish!"

My father beaming.

The man who once took me fishing, and whose blood still courses in my veins, beaming.

At 12:35 the captain said: "Pull 'em up" for the last time, and we headed back to the dock. A short trip. Only four miles to the inlet. Most people onboard were ready to call it a day.

As the mate cleaned the fish of those whose filleting skills fell short of his (and who were willing to pay the one dollar for bluefish, two dollars for fluke), I sauntered up to the kid at the stern. "So," I said to the trip's only representative of his age group. "You gave fishing a six last time I checked. How'd you rate it now?"

"'Bout a four," he said, somewhat overgenerously, I thought, and probably for my benefit. Fact is, in any endeavor, ambition must be met with a level of achievement or there is no gratification. Without fish on the line, words of praise from a captain, or a beaming look from your father, fishing is just getting up early and waiting until you can get home and get on the computer.

Getting to have both bacon and sausage for breakfast does not, all by itself, rate a memory.

Being old enough is hardly worth the effort if it doesn't merit anything.

I sure hope, for a lot of people's sake, that weakies recover soon. When this happens, I hope a four is good enough to warrant a next time.

Chapter 7

After the Storm Passed; Dog Days of Summer

After the storm passed, the tree swallows and bank swallows that had taken shelter in the stands of reed that flanked Dividing Creek erupted from the stalks and dispersed across the marsh, getting back to the serious business of foraging. Before the last heavy raindrops stopped falling, while lightning was still having its will of the eastern sky, the birds were skimming across the marsh at grass-top levels, audibly snapping up insects that had themselves been grounded by the storm.

It was easy to tell the two species apart, despite the lingering gloom and despite the basic similarities between tree and bank swallow. Blue-green above with clean white underparts: tree; brown-backed with a well-defined dark band across the chest: bank.

But on August 2, a novice could master the challenge by noting a single characteristic: the wings. The birds with clean, crisp, symmetrically perfect wings were banks. The ones with gaps or feathers missing along the trailing edges were tree.

The difference has to do with molt. Tree swallows molt their wing feathers in late summer, bank swallows from November through January.

Why? The answer has to do with the different migratory strategies employed by the species. Tree swallows migrate later in the fall—September into early November. In preparation for the journey, they molt their important flight feathers in midsummer.

For bank swallows, the time for preparation was past. Bank swallows were already or very soon would be engaged in a migration that would take them to South America. To do this, they needed their wings in working order. Molting now would make about as much sense as taking your car on a trip when it's halfway through a major repair on the suspension.

In two weeks, it would be easier still to tell tree swallows from bank swallows. At Turkey Point, they'd all be tree. By mid-August, bank swallows are all but gone from the bayshore.

After the storm passed—in fact just about the time a light-ning bolt struck the roof of the Burger King in Cape May Court House, twenty miles to the east, and South Jersey Medical Center lost power—the flock of great egrets that had waited out the storm on the lee side of the phragmites walked or flew the short distance onto the flanking salt hay meadows to resume feeding.

The meadows were flooded. The new-moon tide had sent water spilling out of the creeks and onto the marsh, ferrying with it large numbers of killifish, who had become stranded.

The term *flock* is loosely applied to herons. The egrets were concentrated by opportunity more than by social de-sign. That the dozen or so birds were operating within a dozen long-legged strides of one another was evidence of the number of small fish that had been stranded. If the bounty were less, the distance between the birds would have been greater.

Among the large, yellow-billed birds were several smaller, black-billed ones. These were snowy egrets. Just as white, just as interested in fish as their larger cousins, snowy egrets are more active feeders. Great egrets are stately and deliberate in their movements. Snowy egrets are active, nimble, and more aggressive feeders.

Having different feeding styles reduces competition. It also means that a food type or habitat type not conducive to one species can be utilized by another, further reducing competition.

But even snowy egrets seem sluggish compared with the greater yellowlegs that were charging through their midst. Half the size of snowy egrets, perched on their long bright yellow namesake legs and armed with slightly upturned, stabbing bills, these large shorebirds are, like the egrets, fishermen.

The egrets were local breeders. The yellowlegs were migrants, birds newly arrived from the bogs and wetlands of northern boreal forests. The breeding season over, yellowlegs were heading south—had, in fact, been gathering in local marshes since late June.

Greater yellowlegs, as migrants, are not anomalies. In fact, they are just part of the crowd. Now, in early August, the marshes were filled with southbound shorebirds, almost all adults. In shorebird species, adults migrate first. The young of the year follow.

It's another compensating mechanism. A way of reducing competition, because the greatest competition a species faces is with its own kind. So shorebirds divide their migratory periods between age groups. Older, experienced birds first. Younger birds second. In general, marsh productivity and food availability increase as summer advances. So young, inexperienced birds reach the bay when food availability is at its peak (and after many of the adults have moved on).

What's good for young birds is good for the species. Strategies that work in nature perpetuate themselves, naturally.

After the storm passed, the adult male yellow warbler that had taken shelter in the protective confines of a particularly lush groundsel tree (which happened to be in the bird's territory, which is always used during heavy rain) hopped to one of the outer branches. Took a quick look around. Started preening. Then, surrendering to a sudden urge, threw back his head and belted out a full rendition of his classic song.

Sweet, sweet, sweet, oh-so-Sweet.

It caught the attention of several migrant yellow warblers, whose numbers were close to peaking and whose presence was a result of the storm. Caught aloft ahead of the advancing wall of rain, the birds had put down, just before sunrise, in the island of trees that the resident bird had occupied since April.

The visitors had bred, or were the fledged young of parents who had bred, in willow thickets in the Arctic. They were en route to winter territories in Central and South America, and soon, the resident bird would follow.

Why did the resident bird sing a snatch of song, even though his breeding season was over and his young raised? Even though he hadn't vocalized in almost a week?

Who knows? Maybe he just wanted to celebrate the passing of the storm; or, if you prefer, the bird was stimulated by increasing levels of light as the ink dark clouds that had blanketed the sky moved east.

This is the same as saying, "Because he felt like it."

After the storm, the young great horned owl that had spent the last hour of darkness and the first hour of gathering daylight screaming to be fed finally got the idea that no meal was in the offing. Once this year's offspring had been on the wing nearly two months, the adults had finally severed ties. The young bird had all the anatomical refinement and innate skills necessary to capture its own prey. All it needed now was experience and luck.

Thoroughly sodden and acutely hungry, the bird waited until the rain stopped before leaving its perch on the island of trees, attempting to fly the half mile across open marsh to the neck of land where it had been raised and where it could roost in comfort.

Bad move. Before it could halve the distance, the family of American crows that claim lordship of Turkey Point during daylight hours swarmed all over the youngster, forcing it to take a perch on a blind in the middle of the open marsh, where it could do nothing but duck, repeatedly, as the crows dive-bombed the young owl's head.

The real winner in this bit of drama was the female willet who had been trying to distract the crows from her surviving two young, hidden in the marsh. She and her soon to fledge young were the last willets at Turkey Point. Most of the local birds had departed before the middle of July, heading south in V-shaped flocks for South America. She'd nested late. If other willets had been in the area, they would have rallied at the birds' cries and set up an air umbrella against the crows.

Luck is fickle. This time it favored the willet.

After the storm passed, the roadside-hugging chicory danced in the breeze, their periwinkle-colored petals, still closed, waiting to blossom at the touch of sun. Farther out in the marsh, set like galaxies of pink stars in a verdant green universe, clusters of marsh pinks sued for admirers. Each set of delicate pink petals radiated from a hot yellow core.

It's not just their delicate beauty that makes marsh pinks so standout showy. It's the backdrop of green—uniform in color but not in texture. In places, the marsh looked like some large dog had turned trampling circles in the grass and now left its bed. The "cowlicks" are a signature characteristic of *Spartina patens,* the fine "salt hay" that is the dominant vegetation of high marsh. Each stalk is fitted with a weak spot near its base that is designed to give as the plant matures, allowing the stalk to surrender its vertical stance for a horizontal recline. The weight of one weakening neighbor topples the next, and the grasses fall into swirling patterns that recall plastered-down cowlicks.

The hard rain had accelerated the surrender. The blades that had fallen in the face of the storm would never rise again.

After the storm passed, the small, blackish dragonflies called "salt marsh dragonlets" sat atop the leaves of phragmites, waiting for the sun to warm the air. There were hundreds along the roadside, and, despite their numbers, they were easily overlooked unless you knew to look hard and close.

Not so the larger spot-winged gliders, which would be following on the heels of the approaching cold front that had sparked the thunderstorm. Highly migratory, spot-wingeds catch a ride on the westerly winds associated with mid- and late-summer fronts. Using wind power, the insects migrate hundreds of miles. Going . . . well, nobody really knows where spot-winged gliders go. Like much of life, it's a mystery.

After the storm passed, a very large, better-than-five-foot rat snake set off in search of a temperature setting more to its liking. The rain had taken ten degrees off the biothermometer that is a snake's metabolism. Sluggishly, it made its way onto the road, finding that any residual heat from yesterday's hot August sun, had been leached by the night and the rain.

Drat, it concluded, and it kept going, narrowly avoiding the tires of the car that sped up the road, swerving at the last possible moment.

After the storm passed, I shut off my laptop and called down to Linda that I was heading out to Turkey Point to see if anything was happening. I didn't expect a lot. It was, after all, approaching midmorning, and it was too late in the season for breeding birds to be vocalizing and too early in the season for many birds to be migrating. But my laptop's battery was almost spent, and the thunderstorms being produced by the stalled cold front would keep the battle line between warm, southerly air and cooler air bulging out of the north in the region all day. Why take a chance on zapping the computer?

Sure enough. There wasn't a lot happening. Just the usual. Dog days of summer.

Bank swallows still here. Heard a willet calling (getting pretty late for willet, I thought) and had a yellow warbler do a full song. Late for that, too. Crows were harassing something out in the marsh. Marsh pinks were up.

After about an hour I went home, just missing a rat snake that was crossing the road.

Big one.

CHAPTER 8
Big Red Tomato

DOG DAYS AND SLEEPLESS NIGHTS

Normally I rise before Linda, but a sleepless night, like the one just passed, throws the workings of my internal clock out of whack. The electronic cricket chirps from Linda's alarm came as a shock and an affront.

Swinging my unencumbered feet to the floor (the crumpled sheet was pushed clear to the footboard), I maneuvered a course around panting dogs, who had abandoned their beds in favor of the floor. More by instinct than by design, I found my way down the stairs, into the kitchen, and to a position in front of the coffeemaker.

Ours is a small house, the distance from bed to coffeemaker no more than thirty mostly downhill yards. But the exertion was enough to coat my body with a fine patina of sweat.

Not enough to trickle. Just enough to dampen and cool. Sweating is the human body's primary defense against over-heating. Dogs can't do it (except through the pads of their feet). That's why they pant.

Coffee mug in hand, I made my way to the TV. Collapsed into a sofa whose leather covering stuck to my bare legs. Turned on the Weather Channel. Caught the end of the me-teorologists Jen Carfagno and Nick Walker's tag team act on *First Outlook* and the beginning of the local forecast.

The stats told me what my body already knew: two hours before sunrise, the outside air temperature was eighty de-grees Fahrenheit with humidity levels to match. This would be about average for an Ecuadorian cloud forest at noon. The temperature was projected to climb into the nineties by late afternoon, with humidity keeping pace. It would come close to replicating the environment found in the womb.

It was about this time that I became aware that a cotton-T-shirt-clad apparition that, on the basis of height, outline, habitat, range, and temporal distribution, I concluded was Linda, had appeared in the door. This identification was quickly confirmed by call.

"I. Hate. This," she said, not bothering to explain what "this" might be. It is the vocalization most often made by the Linda Dunne on hot mornings in midsummer. It is uttered in a slow, even monotone that has a sad, plaintive, almost defeated quality.

"Yep," I confirmed. "Gonna be a hot one."

"H.H.H," I added (maybe taunted) when Linda failed to respond. *H.H.H.*, for those of you who have never been condemned to live in the Northeast in July and August, is short for "hazy, hot, and humid." It is not only the most dreaded pronouncement to fall from the lips of TV meteorologists but also the most typical forecast served up in these parts in summer.

Me? I grew up breathing, playing in, and struggling to sleep in this soup.

Linda? She spent her childhood fanned by Pacific Ocean breezes in Southern California and her teenage years up on the prairies of Alberta, where, even in the middle of summer, jackets are standard and the air is so moisture-starved that nosebleed and dehydration are probably ranked numbers one and two on the coroner's checklist under the category "cause of death."

When Linda and I first met, we of course went about the bonding process of getting to know each other's likes and dislikes. Exploring pivotal matters such as favorite ice cream flavor, favorite rock groups, and favorite authors.

At some point, the question of "favorite season" came up, and Linda was shocked by my disclosure.

"Winter," I said.

"Winter?" she repeated, not quite believing it. "Not summer?"

"Winter," I confirmed. "Now that it doesn't mean a reprieve from school, I hate summer."

"You hate summer?" she asked, believing this pronouncement even less. "Summer's the best season of all."

Our courtship began in October. We were married the following July (in California) and spent our honeymoon in the relative comfort of southeastern Arizona. Despite this midsummer break from the Northeast, back in New Jersey, by the end of August, Linda had changed her tune. Summer, after a single East Coast immersion, had fallen to the bottom of her seasonal preference scale. Out of a possible four, it scored a solid five.

"I. Hate. This," she said, again. Then, moving quickly toward the window air conditioner, she hit the On button, bringing ten thousand lifesaving BTUs online. Turned her back to engage the full chilling blast. Opened her arms in the likeness of Christ on the cross. Said something that I couldn't make out over the sound of the fan but if I had to guess I'd say was:

"*Ahhhhh.*"

We don't have central air. Back when ships' blacksmiths were banging boards together, nobody was building homes to accommodate ductwork.

"What are you going to do today?" she asked in tones that had already lost much of their despondency.

"Going to join some migrant farm workers," I said. "They're picking tomatoes, and it's getting close to their last harvest of the season. Want to come?"

"You're crazy," she said. Shaking her head. Closing her eyes. Savoring the life-giving chill.

"What are you going to do?" I asked.

"Well," she said, feigning an indecision she didn't have, "I think for the time being I'm going to stand right here."

"Define *time being*," I said.

"Until October," she said.

I didn't have that luxury. About the time a big, red-hot summer sun would be falling on the fields south of Cedarville, I hoped to be standing on the sidelines, watching the Garden State's first-string pickers harvesting the vegetable that thrives in hot, humid climates and is virtually synonymous with New Jersey and summer.

This is the tomato.

SUMMERTIME. AND THE LIVIN' IS HOT!

Linda's childhood partiality for summer is as well founded as my antipathy for the season. Over much of North America, summer is the time for peak outdoor activity. It's when families travel. It's when family picnics and barbecues and county fairs and no small number of weddings happen.

We are basically warm-climate creatures. Our species evolved from ancestors who lived on the equatorial plains of Africa. Even today, human populations remain largely concentrated around the equatorial middle of the planet and adjacent temperate latitudes. Colder northern and southern regions, lying close to or above the Arctic and Antarctic circles, are relatively unpopulated.

But seasonality and latitude are not the only factors that govern climate, much less weather. Climate is also affected by elevation, ocean currents, and pervasive, long-term weather patterns.

In the Northeast, in summer, it is common for people

to vacation in the mountains or go to the shore to beat the heat.

In New York and Northern New Jersey, vacationing in "the mountains" is virtually synonymous with going to "the Poconos" of northeastern Pennsylvania—unless vacationers elect to go, instead, to the Catskill Mountains of southeastern New York, which are geological kin. Why? Proximity combined with elevation. The Pocono Mountains, less than two hours' drive from Manhattan and lying at about two thousand feet, are cooled by both altitude and their vegetative cover. Air temperature falls three degrees for every thousand-foot rise in elevation. Even without their solar-energy-deflecting forest cover, the Poconos are approximately seven degrees cooler on any given day than New York City (which lies mostly at sea level).

Seven degrees can mean the difference between eighty-three degrees Fahrenheit (which Linda and I consider the upper limits of human tolerance) and ninety degrees Fahrenheit (which we have no problem calling "hot").

In fact, ninety degrees Fahrenheit constitutes the National Weather Service's official threshold for "hot." Four days of ninety-degree temperatures and above qualify as a heat wave. And *heat wave* in the Northeast, in summer, is a double four-letter word.

It sends people scampering to municipal pools, air-conditioned malls, and, best and most of all, "the Shore," where cooler ocean breezes keep summer heat at bay.

Actually, this is somewhat misstated. A sea breeze doesn't

hold warmer air in check. It supplants it. The process works like this.

What we think of as the common, daily shore wind pattern results when the air overlying the land heats more rapidly than the air overlying the adjacent ocean waters. This land-heated air expands and rises. Cooler, denser air overlying the water flows landward to fill the vacuum. The result? A cooling breeze, coming in off the ocean, at a pleasant speed of ten to fifteen miles per hour, just when we comfort-seeking humans need it.

At night, the process and the wind direction are reversed. Land cools more quickly than water. It's the air over the ocean that rises, causing the air overlying the land to flow to the coast.

Of course, a coastal sea breeze is a highly localized phenomenon. You don't have to go far inland before it loses its punch (Linda and I, a mere four miles from Delaware Bay, are mostly beyond its friendly reach). Our weather—in fact, the summer weather enjoyed (or suffered) by most of the eastern United States—is controlled by a huge summer pattern dominated by the Bermuda High—a large, stable high-pressure air mass that moves north out of tropical regions in the summer, parking itself in the Atlantic Ocean near the island of Bermuda, about seven hundred miles off the coast of North Carolina.

The Bermuda High, and its associated low-pressure system anchored over the hot interior United States, combine to produce a steady and pervasive summer airflow that comes up from the south and streams north across the eastern United States. Already tropical in origin and nature (meaning warm and moist), this southerly airflow picks up additional heat as it passes over the sun-cooked southern United States. By the time these heat-fed winds get to New Jersey, they may reach and exceed one hundred degrees Fahrenheit.

The record for Cumberland County, a mark reached in both July and August, was a sultry 104 degrees Fahrenheit.

With the summer jet stream commonly stuck farther north, preventing cooler air from reaching the region, this summer pattern of hot, southerly air dominates weather and causes unrelenting heat, broken only by the mercy of afternoon thunderstorms.

I didn't know a damn thing about Bermuda Highs when I was a kid. I just knew that summer, in New Jersey, was miserably hot.

We didn't have air conditioning when I was growing up. No, the ductwork was fine. We were just too poor to afford the luxury.

In the summer, during the hottest part of the day, I'd retreat to the coolest corner of our basement and read. At night, lying naked and sprawled atop beds whose sheets were hot to the touch, my brothers and I would struggle to sleep. We'd turn our pillows frequently to feel the momentary relief of the sides not warmed by our heads. We'd viciously kick away any fold of sheet or shed any pajama bottom that fell against bare skin.

What I remember most about those nights was the sound of our dog, Troubles, panting. The hum of the attic fan that did nothing but draw outside air, as hot as inside air, in through the window. The sound of the clock striking hours never heard at any other time of year—eleven ... twelve ... one ... And the grating calls of katydids—arboreal crickets whose monotonous mantra—*kay-tee-did* ... *kay-tee-did* ...—went on and on till dawn.

Even now, decades later, the sound of a katydid makes me feel ten degrees hotter than the thermometer reads.

Bear in mind that I grew up in North Jersey, where, according to the laws of altitude and latitude, it should have been cooler. Small degrees, small matter. When people ask me what I remember most about my childhood summers in New Jersey, the very first thing that comes to mind is the heat. It is branded on my psyche.

But the next thing is tomatoes. People may wilt under a New Jersey summer. But New Jersey's state "vegetable" thrives.

THE GARDEN STATE?
Whether you travel to, or from, the shore on Route 47, you will pass a handful of roadside enterprises called "farm stands"—vegetable markets that feature the produce in season. They open their doors when asparagus emerges in April—big, succulent, two-fisted bundles offered at irresistible prices.

The season continues, for the larger stands, until about Thanksgiving, when the demand for decorative gourds and pumpkins ends and frost blights the last of the broccoli and cabbage.

In between you'll find a cornucopia of produce bearing the Jersey Fresh brand: rhubarb, lettuce, strawberries, onions, squash, blueberries, cantaloupes, corn, watermelons, eggplants, peaches, cherries, plums, pears, apples . . .

And, of course, tomatoes. New Jersey's official state vegetable.

"Jersey Fresh," by the way, is a marketing campaign concocted by the state's Department of Agriculture. It assures market-hardened consumers that what they are buying at local farm stands is, in fact, locally grown. It is also an affirmation of our state's proudest boast and New Jersey's nickname, the Garden State.

Etched on the bottoms of automobile licenses, disbelieved by people whose image of New Jersey was forged by a trip up the New Jersey Turnpike, it is nevertheless apt. Since pre-Revolutionary days, New Jersey has served as one big truck farm for the populations of Philadelphia and New York. With thousands of acres under the plow, a growing season exceeding 190 frost-free days (in the south), and proximal access to 60 million customers, it would be a wonder if New Jersey were not an agricultural powerhouse.

New Jersey is the place where Clarence Birdseye developed the freezing and packaging process that revolutionized the food industry. It is where, in 1897, Joseph Campbell exploited the principle of condensation that made his name (and soup) an American icon (the company headquarters remains, to this day, in Camden, New Jersey) and where, in the first half of the twentieth century, when faced with the growing competitive might of large retail food market chains, a grass-roots marketing backlash evolved.

Small farmers who were being muscled out of the produce market realized that, in a crowded drive-through state, they didn't have to sell their produce at small margin at auction. The consumer market was coming to them, driving right by their doors.

They set up roadside stands. Offered local produce. Left cigar boxes and coffee cans out for people to deposit exact change (and I'm delighted to tell you that in A.D. 2009 you can still find successful unmanned farm stands, in Cumberland County, that do business by the honor system).

Among these marketing pioneers was the Buganski family, owners and operators of Lillian's Market. First located on Route 47, near the village of Delmont. Relocated in 1980 east of Port Elizabeth, Lillian's is homegrown, third-generation-run, and the place where Linda and I buy virtually all our produce despite proximity to three other farm stands (two manned, one honor system) that are about equally close.

I'd rather change barbers, auto mechanics, or dentists than buy my produce anyplace but Lillian's. And Lillian's customers, some of whom are second- and third-generation themselves, feel likewise. Among them you'll find some of the pickiest tomato buyers this side of the Philadelphia market.

Ask a roomful of New Jersey natives to name *the* vegetable that is synonymous with their state, and, almost to a man, woman, and child they will say "tomato." They will also be wrong. Not because of partiality or association but because of classification. The tomato is not, botanically speaking, a vegetable. The tomato is a vessel of seeds surrounded by a fleshy womb. It is a fruit; not that anybody, except maybe the federal government, cares to argue the point.

In very fact, the question regarding the classification of *Solanum lycopersicum* (literally, wolf peach) was, ultimately, decided not by Linnaeus, who named the plant, or botanists, who later placed it in the genus *Lycopersicon*, but by the United States Supreme Court. In 1893, in a ruling on a case to determine whether the tomato was a vegetable, thus subject to tariffs imposed in 1883, or a fruit, therefore tax-exempt, the highest court in the land declared that this taxonomically certain berry was in legal reality a vegetable.

As the foundation of its decision, the court concluded that, insofar as the tomato was properly served as part of a dinner's main course and not dessert (as would befit a berry), the fruit must be in league with beans, beets, corn, and other items labeled (and taxed) as "vegetables."

The real point here is not to challenge convention or taxonomy. It is that when I decided to write a chapter about this seemingly simple subject, I had no idea I was opening such a can of tomatoes! The story of this vegetable/fruit is so deliciously multifaceted that I came darn close to scrapping the whole project and just writing a book about the tomato.

Until I discovered that it had already been done, and admirably so, by Andrew F. Smith, author of *The Tomato in America*.

A Tomato Runs Through It

Like those other farm market favorites, corn and squash, the tomato is an all-American fruit—providing you take a hemispheric, not a U.S.-centric perspective. As postulated by Andrew Smith, the genetic ancestors of the modern tomato heralded from the Andes Mountains of South America—about where Peru is today. Never cultivated or, apparently, eaten by the Incas, the fruit in some undetermined fashion made its way to Central America, where the *xitomatl*, as it was known, was domesticated and became part of the diet of, among other people, the Aztecs of Mexico.

Shortly after Cortés's conquest of the Aztecs, the plant found its way to Europe. By the mid-1550s the "golden apple" had secured a place in Italian cuisine (where it was consumed with oil, salt, and pepper) and in Spain (where it was known by the name *pomme de Maure*, Moor's apple).

It was, apparently, a misinterpretation of the Spanish name that resulted in the popular renaming of the plant "love apple." Phonetically speaking, *pomme d'amour* (apple of love) is pretty close to *pomme de Maure*. In keeping with the popular new trade name, the tomato was considered in some quarters (most notably France) to have aphrodisiac properties.

Despite its New World origin, the love apple was not an instant hit with the American market. In fact, early American colonists appear to have regarded the plant with disdain and trepidation.

Maybe it was the suspicion that the lush red fruit would spur licentious behavior—a freedom of expression not favored in Puritanical society. More likely the antipathy was rooted in the plant's resemblance to, and botanical kinship with, deadly nightshade, a hallucinogen, in whose family, Solanaceae, the tomato is found (along with potatoes, eggplants, and tobacco).

Nightshade was one of the standard apothecary items of practicing witches. When it was administered vaginally with, for instance, a handy broomstick, the minds of users took flight. Witchcraft was no more tolerated by colonial Puritans than promiscuity, so the tomato enjoyed double-hex status in early American society.

But not by all. Thomas Jefferson, among others, cultivated (and presumably ate) tomatoes at Monticello. In 1820, a brave tomato firster named Colonel Robert Gibbon Johnson is alleged to have announced that, at high noon on September 26, he would eat a quantity of tomatoes in front of the courthouse of Salem, New Jersey.

Some accounts say the courthouse was in Salem, Massachusetts; some Boston. It hardly matters. The popular story is unsupported by any evidence. Wherever the historic event is alleged to have taken place, a reported two thousand people gathered to watch the colonel die in agony, which, to their compounded disappointment, he did not.

I actually read an account of this apocryphal story in one of my grammar school reading books (where it was presented as fact). Years later, I am brought to question its authenticity. It seems highly unlikely that, in 1820, two thousand people would have had the latitude to interrupt their work, during the harvest season, just to watch a man die. This, after all, was an age when hangings were public and child mortality rates made death almost as familiar as birth. One man's public suicide attempt couldn't have been that interesting, could it?

Whether Colonel Johnson's demonstration did or did not happen, or had anything to do with the turnaround of the tomato's popularity, is unclear. What is certain is that, shortly after the Civil War, the tomato grew in acclaim. Today, it is not only the nation's favorite vegetable but also, as stated, New Jersey's official state vegetable.

You're talkin' tomato, you're talkin' state pride. Tomatoes are arguably but by no means certifiably, among the few things New Jersey residents can be universally proud of.

Our corn's pretty darn good, too.

In Search of the Jersey Tomato

My father, Gerald W. Dunne, was a reluctant gardener. No doubt he inherited his brown thumb from his father, Edward F. Dunne III, who argued convincingly and with foundation that planting a garden made not a lick of sense. "Once things are ripe," the practicing patent attorney argued, "vegetables are so inexpensive you can hardly afford not to buy them."

This conviction did not prevent my grandfather from cultivating yeast, in vats of hops-enhanced barley, in his basement during Prohibition.

Anyway, and as stated, my father was not an enthusiastic gardener. Nevertheless, every spring he dutifully dug and planted a garden in the weedy back corner of our property. He planted carrots, which were always stunted. He planted zucchini, which, unattended and incubated by New Jersey's summer heat, grew as big (and about as succulent) as the 155 howitzer shells he and the other members of his outfit directed at German positions during "the war."

And he planted tomatoes, which, unlike the other vegetables, he tended religiously and harvested with obvious pride.

Even slapped between the nutritionally bankrupt stuff that passed for bread in those days, a tomato sandwich was a summertime treat in household Dunne. Cut thick and double-layered. Slathered in store-brand mayo and garnished with salt and pepper. There was nothing in the world that tasted better than a sandwich made with tomatoes straight from the garden and crafted by my father's (or mother's) hand.

I enjoyed mine with milk.

He had his with a beer.

His father's son.

Choice of accompanying beverage notwithstanding, almost every spawn of a New Jersey suburb has memories similar to mine. Jersey tomatoes, fresh off the vine. Their juice overflowing the cutting board. Their sweet and acid taste making your taste buds flip-flop like poll-driven politicians in a tight race.

Even now, as I am just writing these words, my salivary glands have kicked into overdrive and I find myself ignoring the cereal still lodged in my tummy and looking at my watch to see how long it is before I can go downstairs and put a rich red, grapefruit-size tomato under the knife.

But will it be, in this day and age, a real New Jersey tomato? The tomato of my youth?

I posed my question to David Shepherd, who, along with his brothers Tom and Erwin, runs Shepherd Farm—one of Cumberland County's agricultural powerhouses, located just south of Cedarville. The three brothers, graduates of the Cornell College of Agriculture all, are tenth-generation Cumberland County farmers, the ancestors of five brothers who emigrated to Back Neck, in "West Jersey," in 1683, following Lord Cromwell's Protestant victory during the English Civil War.

They fought on the winning side.

The Shepherds own the trademark *"The* Jersey Tomato," so what they grow, and what they ship to several well-known supermarket chains operating from New England to Ohio to Virginia (and, of course, New Jersey), is, at the very least, official. So what distinguishes a Jersey tomato?

"A Jersey tomato is all red. Real solid. More juicy. High in sugars. High acid. Better tasting," said the wide-faced, white-haired, bespeckled, and beaming member of the Shepherd trio. At the moment, he was trapped by my interview behind his desk, but his workplace is more commonly, and more to his liking, outdoors in the field.

He looked like a businessman. He was a businessman. And why not? Agriculture is big business. A $700 million per year business in New Jersey.

"I think a lot of it has to do with the soil type," he added after a moment's reflection. "A tomato grown in New Jersey just tastes like a Jersey tomato."

While not the only crop the Shepherd family grows, tomatoes are a big part of their business, their image, and their pride. They currently have twenty acres in tomatoes. From these intensively managed acres, they will harvest approximately 1 million pounds of tomatoes. Admittedly, this is only a fraction of the 11 million tons of tomatoes produced annually in the United States. Despite its cultural affinity, and measured solely on quantity (not quality), New Jersey is not even ranked among the top tomato-producing states. But if you have shopped in a New York, Pennsylvania, or New Jersey supermarket, chances are good that you've enjoyed a Shepherd-grown Jersey Tomato.

"What variety are you planting?" I asked, getting right to the meat of the matter. There are scores, maybe hundreds, of tomato types. They vary in size, shape, color, texture, seasonality, acidity, and sweetness, to mention just a few, most obvious, qualities.

Many of the most common and popular commercial varieties are hybrids—hardy, prolific producers bred to meet the practical needs of growers as well as the epicurean favor of consumers. While this is not specifically why the plants are known as hybrids, for all practical purposes it could be.

Then there are heirlooms. Older strains, many with a regional flavor, commonly favored by backyard gardeners and grown more for taste than for portability or marketability.

"We're growin' Indy," Dave replied. It wasn't a variety I was familiar with.

"Why?"

"Best flavor, best to ship. You need to select for shipability," he asserted, showing his businessman's side. Also, I found out, the Indy is an early-ripening tomato. At Shepherd they're harvesting tomatoes from the end of June to early August.

Why not later? Because by late summer local farm stands are overflowing with tomatoes. Prices plummet. In farming, to make money, you have to reach the market with the product people want before everyone else does. Earlier in the 2008 growing season, when the market for East Coast tomatoes was bolstered by a nationwide salmonella scare (which was later traced to peppers shipped from Mexico), the price for tomatoes was up around twenty dollars for a (approximately) twenty-five-pound box.

Businessmen, whose business is tomatoes, love twenty dollars a box.

By August, prices were down to around twelve dollars per box. Given the costs of production, that's breakeven.

Businessmen, who want to stay in business, aren't interested in breaking even.

"How come you don't raise the New Ramapo?" I asked, naming a tomato touted by the highly respected Rutgers School of Environmental and Biological Sciences as *the* varietal reincarnation of the venerable Jersey tomato, the tomato of my youth.

"Doesn't ship well," said Dave Shepherd, businessman and Cornell graduate.

Did I mention that the color combination for the well-known Campbell's soup can was chosen to match the school colors of Cornell? Thought not.

HOME (STAND) GROWN

"Sun Brights," Tom Buganski said, grabbing a large, red-ripe tomato off the top of one of the baskets perched on the gate of the pickup. Rubbing a grime-coated thumb over the stem side of the fruit. Noting the scarlike blemish and not liking what he saw. "They're the biggest and the best," he asserted.

Apparently Tom wasn't a Rutgers alumnus either.

"Hey you, Blanca!" he shouted toward the sturdy Mexican woman working the rows nearby. "No es _____ Farms," he said, holding the blemished fruit aloft (and castigating one of his large commercial rivals).

Blanca seemed not to take the criticism very seriously, recognizing, as I did, that it was being made mostly for my benefit. She listened politely, then bent over and continued the job of filling another basket, already brimming with bright, red, lush, vine-ripened fruit.

There was nothing wrong with the tomato in Tom's hand. The healed cut was, after all, only skin deep. But Tom's point was apt. People who buy their tomatoes at farm stands (like Lillian's) tend to be choosy, and they also tend to be regulars. As anyone who operates a farm stand will tell you, tomatoes with blemishes are selected against, and farm stands that offer less than visually appealing produce are, too.

"Discard them here, or discard them there," Tom said. "People don't want 'em and if we bring them to the stand we'll just have to throw them out."

Almost all of the tomatoes picked by Tom and his crew are destined for sale at farm stands—their own family stand or one of the smaller roadside stands they sell to. Only a small percentage, the surplus, go to the farmers' auction block in Vineland.

Since they are going into the hands of consumers hours after they are picked, the fruits are genuinely, not legally or technically, "vine-ripened." Tomatoes on the large commercial farms, like Shepherds', are picked mostly green because they are harder and more resilient then, so they handle and ship better. All it takes is a little color, or "blush," on the bottom of the fruit for a tomato to legally qualify as "vine-ripened." While they are not "gassed" (artificially stimulated to ripen by being subjected to ethylene gas), the final stage of ripening for green-picked fruit is off the vine.

The fruit being picked by Tom's crew was eat-off-the-vine ripe. It requires more care and it takes more time to grow tomatoes this way. And as anyone who has ever bought tomatoes from a farm stand will tell you, they just plain taste better.

There's another reason farm stand tomatoes are commonly more flavorful than those found in the stores. Because they are not subject to extensive shipping and handling, tomato varieties grown for local farm stands can be selected with an accent on flavor rather than sturdiness.

Of course, since the tomatoes are being picked fully ripe, the people doing the picking have to exercise an extra measure of tender, loving care. Hence Tom's tutorial oversight.

"Don't know what I'm sayin' to them most of the time," he confessed, smiling ruefully. "It comes out Italian, mostly, whenever I try speaking Spanish."

"How long have they been picking for you?" I asked.

"Lionel and Blanca, four years. See here?" he said. "This is what I'm talkin' about," he said, reverting to a conversation we'd been having about the difference between "determinate" and "indeterminate" tomatoes a couple of subjects back.

I was beginning to get used to the wonderfully eclectic workings of the sixty-two-year-old farmer's mind. Lillian's son, Tom, was the actual farmer in the family—the guy who grew the produce that shoppers were buying at Lillian's.

Have you ever wondered whether the vegetables found in farm stands are really and truly grown on local farms? I certainly have. I can tell you for a fact that at least at Lillian's the produce is not only farm-grown but grown on their own farm—250 acres bordered by East Creek in Eldora, of which 60 are planted by Tom and his Mexican American crew.

I had arranged to meet Tom to watch the operation. Despite the heat, an afternoon rendezvous was set. "We pick in the afternoon, after the sun has dried the dew off the tomatoes," Lillian explained. "Be picking Thursday, if that suits you."

It did. And while Lionel and Blanca filled their buckets and loaded the van, Tom filled my ear with the lore and laments of those whose lives are indentured to the earth and whose fortunes are governed by market forces and weather—neither of which is in their control.

Tom, a broad-faced, compact man, was wearing a New Jersey Fish and Wildlife cap and clothes that might have been denim beneath their honest layer of soil. At first glance, he looks Irish—a classic twinkly-eyed mick whose immigrant ranks were regarded with suspicion and disdain in the middle of the nineteenth century. Back when signs like IRISH NEED NOT APPLY were snobbishly fashionable among those whose ancestors had stepped off the boat a few decades earlier.

My ancestors, actually, on my father's side. His mother's father, my great-grandfather Powers, was an Irish immigrant lad who began his career as an usher in a Chicago theater. He ended up founding the Powers Theater chain. In the spirit of America, a self-taught, self-made man.

The Dunne side of my father's clan trace their lineage back to an Irish patriot, Edward F., who is alleged to have fled to the United States after having taken a failed shot at a British soldier (according to family legend it was drink that spoiled his aim). He went on to make several million dollars, first in the brick industry, later distilling whiskey. Most of his earnings were confiscated by the government of his adopted country when it discovered they were being used to bankroll a planned invasion of Canada by an army of Irish patriots.

A patriot to the end, Edward F. might nevertheless have lapsed into obscurity except for one thing. His son, my great-grandfather, went on to become governor of Illinois.

But Tom's last name gives the first clue, and if you look closer, you see his face is stamped with the broad-featured openness that is so classic of the Poles—another hard-working, opportunity-seeking, and in their time greatly derided immigrant group. In the late 1800s, many of these Polish immigrants settled in the anthracite coal region of northeastern Pennsylvania, where they and their sons went down in the mines to do the hard, dangerous, but necessary work that many native-born Americans shunned.

My ancestors on my mother's side worked in those mines. When I was a child, my grandfather Peter Olshevski could enthrall us for hours with stories about his adventures belowground, where he worked as a young man.

I never personally suffered any rebuff for being half Polish. Perhaps I was shielded by my Irish surname. But my Polish uncles knew the sting of affront, and some, who derided them, felt the backlash. My uncle Pete, whose love of the outdoors I inherited along with his Browning Auto-5 shotgun, actually gave up going to bars in the interest of maintaining ethnic harmony.

He was only five foot six, but his lack of stature sometimes caused careless men to underestimate a physical strength that a look at his arms and shoulders would have confirmed. He also possessed a quickness that was frightening to behold. Pete was not a violent man, but he did have a refined sense of pride and a low threshold for insult, and it became clear that, unless he steered clear of bars, some stupid, beer-enhanced *Mayflower* descendant was going to get seriously hurt.

He died duck hunting at the age of thirty-nine. Massive coronary, his second (at least that the family knew about). They found him lying in front of his blind with his Belgian Browning lying beneath him and his dog, Pan Tadeysz, standing guard over him. Translated from Polish, Pan Tadeysz is Sir Theodore. Theodore was a warrior saint.

Then it was waves of Italians who faced the immigrant hazing America doles out to its latest ethnic arrivals. In the first half of the twentieth century, the Italians came out of Philadelphia by the hundreds to work South Jersey's fields in summer, returning to the city after the harvest. Growing as fond of the region and the opportunity as they were weary of the seasonal commute, many ultimately stayed, becoming part of the large Italian population that is endemic to the agricultural hub that is Vineland.

In the latter half of the twentieth century, by the time Tom was helping his father with the operation of the farm, they were hiring Puerto Ricans as summer help. Now, virtually all of the seasonal farm labor in South Jersey (and else-where across North America) comes from Mexico.

Young males, mostly in their late teens and twenties, work year-round, following the crops north with the summer season and retreating to work in the fields of Florida and Texas in winter. They work six-day weeks, ten-hour days, send most of their money home to their families, and plan to return to Mexico in three years. Couples like Lionel and Blanca, Mexican natives but Vineland residents, are the exception.

WHERE THE FINGERS MEET THE FRUIT

Earlier in the week, I'd visited one of the immigrant camps. My grasp of Spanish is rudimentary, so I was accompanied by a friend, Erika Peterson, who works for a nonprofit health clinic serving the needs of migrant farm workers in Kiptopeke, Virginia.

The camp, bracketed by farmland, consisted of two one-story barracks constructed of cinder block. All the camps, of which there are a dozen in the area, have thought-provoking names (for example, the Camp of the Spiders).

The workers share showers and a common kitchen, and had just finished their day when we arrived. We caught them between cleanup and dinner, and more or less unaware. Needless to say, our intrusion was met with a certain guardedness.

But I was keen to learn more about the people who are the primary link between the soil that grows and the consumers who ultimately buy the crop that the "Garden State" is famous for. The people whose hunched forms I've seen so many times through the windows of my car but have never had cause—or maybe courage—to engage. The people who constitute America's newest wave of immigrants and who have, like all those nationalities before, come to the fields of South Jersey in hopes of bartering hard work for a better life, bridging the gap that lies between America's wants and its neglects.

Despite their reserve and the ten hours they'd spent in the fields, the workers finally opened up a little, won over by Erika's fluent Spanish and easy empathy (her current boyfriend was a Mexican farm worker).

The fact that Erika is also strikingly feminine didn't hurt our effort.

"Tomatoes," I learned from these experts, "are the hardest crop to harvest." The baskets are heavy (about forty pounds full). They must be moved along as the pickers move, so it is not simply a matter of picking up a basket once and bringing it to the waiting trailer. Also, the tomatoes are hidden among the leaves. The more the rows are worked, the less fruit is picked for the effort.

The workers are in the field, picking, at seven. They take a break at ten. Get an hour off at noon. Break again at three. Work until six or seven. Load up and return to their barracks.

They work as teams of twenty-four. They are paid by the weight of their day's total harvest multiplied by whatever the farm is paying, divided by twenty-four.

On average, a worker makes about $350.00 per week.

Calculated at a sixty-hour workweek, that comes to about $5.80 an hour. It's not much. But it is considerably better than the wages they would be making in Mexico doing the same work. Most, and perhaps all, of the men on the twenty-four-man team we interviewed were from the region around Veracruz. Ranging in age from late teens to late forties, all had worked as farm laborers in their native Mexico before coming to the United States.

I asked them what it was they would want me to convey to you, an American consumer who is the beneficiary of their labor. I was told that what they want most is to be allowed to work and enjoy the benefits that go with this.

You know: the American dream.

What I read in their faces and heard in the timbre of their voices was an immense pride in their vigor and their ability to do the hard work shunned by others, and, unless I'm reading too much into this, I noted, too, an overt affinity for their country of employment. There were, among them, far more I ♥ USA T-shirts and New York Yankees baseball caps than might be encountered in a random sampling of native Americans.

Particularly any sampling drawn from South Jersey! This is, after all, Philadelphia Phillies territory.

"How many of the men we spoke to are illegals?" I asked Erika after we'd left the camp.

She looked at me like I was daft. "All of them," she said. And while this was sinking in, she added, "Remember, too, that all you saw were men in the camp and that they all came from the same region. Somewhere there are villages that are all women and children who are waiting for their men to return."

"Will they?" I asked.

"Most," she said, smiling.

"The workers have a saying," she added. "*Una vez que probo norte, siempre quiere volver.* Once you've tasted the North, you will always want to go back."

To be honest, and for the record, I don't know that the workers I spoke to were "illegals." We didn't ask, and who is going to tell? Their nervousness certainly stemmed, in part, from having two strange Anglos show up at their door virtually unannounced. Their reluctance to have their photos taken, understandable modesty.

One thing I am sure of: you don't have tomatoes without tomato pickers. And one thing I would put money on: someday, the sons and daughters of some of the men whose backs are bobbing among the vines of Cumberland County today will be driving to second homes and condos at the shore and stopping at farm stands along the way.

Searching, just as their immigrant forebears did, for the perfect Jersey tomato.

ASK THE EXPERTS

Despite all I'd learned and read, I still wasn't certain I understood what it meant to be a Jersey tomato. So after Lionel and Blanca finished loading Tom Buganski's pickup with the day's fresh crop, I followed Tom back to Lillian's, figured that I'd go get the last word on taste right from the consumer's mouth.

It was Thursday afternoon, and business was brisk. Eighty-year-old Lillian was ensconced behind the counter, greeting customers with a smile and sometimes a name. Her daughter Pat was shuttling produce forward from the back to the fast-emptying bins.

Despite arthritic hands, Lillian moves behind her counter with the deftness of a casino dealer behind a black-jack table. Despite her ten-hour-a-day, seven-day-a-week schedule, Pat hustles with the bustle of a preschooler.

Nodding to both of them, I took a position near the open bin of lush red tomatoes. There for the picking, at a mere $1.49 a pound or, if you preferred, $10.00 a basket, was the succulent essence of summer.

Almost everyone who stops at a farm market this time of year buys tomatoes. If they don't start with tomatoes, they finish there—after surrendering to the temptation of corn at $4.00 a baker's dozen, peaches at $1.79 a pound or $8.00 a basket, blackberries at $2.00 a pint.

My first expert was a slender woman in her late forties driving a silver Subaru. "Bound for the shore," I learned. Picking up tomatoes for salsa for "the party," she confided.

She dawdled over the selection, finally picking four big, ripe tomatoes of about equal size—then a pint of cherry tomatoes.

"Why the cherry tomatoes?"

"Pop them as I drive," she said with gusto.

A real New Jersey tomato maven.

The second expert was a jewelry-encrusted gentleman who resembled a fleshy Groucho Marx and who, with some effort, levered himself out of a black Cadillac so shiny you could probably have seen the glint of it from the International Space Station. He chose a single tomato—the biggest one in the bin, near as I could tell—and headed straight for the checkout counter, buying nothing else.

"I like 'em firm, not soft. You know what I mean?"

"What are you planning to do with it?" I asked.

"White bread, salt, mayo," he said in gastronomic shorthand. "Who could ask for anything more?" I couldn't immediately think of anything (except perhaps pepper).

Expert number three was exactly that—a trio of experts that climbed out of a white Chevy Tahoe. The team consisted of Mom and two experts in training, ages two and five. Despite the challenge of mentoring and monitoring, Mom's selections seemed quick and sure.

"So what's your technique?" I said. "How do you know which ones to choose?"

"Well, first," she noted wryly, "I pick the ones my kids have bruised. Then I look for bright red color and note whether there's any green left on the top or bruises on the bottom."

"You've done this before," I said.

"I spent my teenage years working in a farm stand," she admitted. "Up on Route 422."

"Not one of Rossi's girls?" I asked, naming a rival stand whose marketing strategy includes multiple roadside signs touting the feminine quality of its vendors.

"Yep," she said, beaming and removing a tomato from

the hands of her most junior understudy, examining it for quality (and wear), and returning it to the bin.

My final expert was a woman in her fifties. She was brisk in movement, brusque in manner. She grabbed a bucket of tomatoes (no picking, no fanfare), put them on the counter, then in short order grabbed a dozen ears of corn and an armload of squash. Paid cash and started for her car.

"You're the only person to pick up a bucket," I noted. "Everyone else went for the singles."

"I'm on my way home to Pennsylvania," she explained. "Used to live here. Visiting friends. My marching orders from home were 'Be sure to pick up tomatoes.'

"There isn't much that you can say that's nice about New Jersey," she added, "but it sure does have the best tomatoes."

That's true, even if nobody seems able to define what a Jersey tomato is. Maybe, like memories of childhood, the notion of the real Jersey tomato is just, in the final analysis, a state of mind.

CHAPTER 9
Catching the Comet's Tail

Turning off the light, opening and closing the kitchen door as quietly as a child not wanting to be heard by adults, I stepped into the night and peered into the sky. Hoping the weather forecast was right. Hoping my 2:00 A.M. wake-up hadn't been in vain.

It took a moment for my eyes to gain their "night vision." A moment more to find a pinprick of light strong enough to penetrate the milky veil drawn over Mauricetown—but I succeeded.

There, high overhead (the only wedge of sky not blocked by buildings, trees, and star-murdering streetlights) were several stars, bright enough to defeat the localized light pollution that is hated by star watchers the world over.

That's why I was heading for Turkey Point, end of the universe, New Jersey. More than this, I was going there with a special objective in mind. I was going to catch a comet's tail.

The date was August 12. Prime time to catch the Perseid meteor shower—the most watched event in the summer sky.

THE END OF THE ROAD

Coming to a stop just before the footbridge that marks the end of road, being careful to keep my eyes from the sky, I grabbed my binoculars and a flashlight, making sure that the one I grabbed had the red filter.

Why a red filter? To keep from losing my night vision in the event I turned on the light. White light, even white light directed away from your face, temporarily undermines your ability to see clearly in the dark.

Why my careful avoidance of the night sky? Superstition. I didn't want to see my first shooting star before I could make a wish. Kills the magic.

After moving the car's interior light switch to off, I opened the door and stepped out into the universe.

It was a beautiful morning, providing you take the broad interpretation and agree that anything that falls after midnight and before noon constitutes morning. Cool, dry, with a modest breeze coming from the west.

And directly overhead was a gorgeous sky. Clear, cloudless, and dark except for the millions of stars stropped to a sharp-edged luster by the air moving in from Canada, courtesy of an early-season cold front. I couldn't have asked for better viewing conditions (and in the days leading up to the Perseids, I asked often).

Warm, moisture-laden air oozing up from the south dampens our view of the sky by increasing the scattering effect of light. Dry air reduces the scattering of light, so stars shine brighter. In fact, conditions were so good that the hazy band of light that is the edge of our galaxy was, for the first time in many hazy, hot, and humid days, plainly visible overhead. Known to all, seen by increasingly few, the Milky Way is a wheel of stars that appears—from the perspective of our planet, stuck as we are out on the edge of that celestial pinwheel—to be a stripe, a milky (high)way bisecting the sky.

When I first moved to Mauricetown, seeing the Milky Way was an every uncloudy night matter of fact. Now, the galaxy we live in is visible only on the clearest nights. Such nights are most common in winter, much less so in summer.

I was so giddy at my good fortune that I almost forgot to make a wish. Lucky for future residents of Cumberland County, New Jersey, I remembered in time.

Just in time, because to the northeast, out of the corner of my eye, I picked up an eyelash-thin flash of bright white light.

Before I could think, much less say "Awww" to register my disappointment (you always want your first shooting star to be a big, bright, streaking buster of a meteor), a second, larger meteor flashed into view following the same trajectory as the first, leaving a white vapor trail in its wake.

Obviously, that was the one I'd pinned my wish on. The first was just a scout.

But whether they were paired by physics, fate, or fancy, it was still difficult to believe that, despite their inspirational merits, all I had really witnessed were the collisions of tiny, sand-size particles of comet-spawned junk smacking into the earth's atmosphere and vaporizing in the heat of impact.

Because that is all a shooting star is.

Don't you just hate it when magic is reduced by science to plain old phenomena?

The undermining truth didn't matter. I said "wow" and, despite the fifty-nine-degree temperature, was warmed by the thought that my wish was as good as got. The temperature at the shooting star's point of impact was a good deal hotter. About three thousand degrees Fahrenheit.

I hoped that in other places other people were getting their wishes, too. I looked at my watch. Saw the little hand on three, the big hand on twelve. Noted, with glee, that there were still at least two good hours of celestial fireworks ahead of me. Sunrise was 6:10. But the first magic-nullifying glow of morning light would begin to infiltrate the sky before 5:00.

With no small measure of satisfaction (but no surprise), I also noted that I was the only person on Turkey Point Road. Sky watchers, in A.D. 2008, seem to be diminishing as fast as the stars.

It was about a mile of mostly open road back to the

woods bordering the marsh. I set off with my eyes glued to the sky and my feet finding their way.

Particles Colliding in the Night

The Perseid meteor shower might be spectacular, but it is not unique. Our planet is continuously bombarded by particles of matter that originate in space. In fact, at almost any time of year, observers can see between one and three shooting stars per hour given good viewing conditions and an auspicious, unobstructed view of the sky.

These chunks of space junk, called "meteoroids," range in size from a grain of sand to a two-story house. Matter smaller than this is called "interplanetary dust." Objects larger than this are known as "asteroids." Composed mostly of metal or rock, these pieces of matter are, like our Earth, in orbit around the sun. Entering the earth's atmosphere at speeds that may exceed forty miles per second, meteoroids are transformed, by heat, into visible events called "meteors." The streaks of light that we see are the byproducts of matter striking matter at high speed.

When we watch a shooting star, we aren't really seeing the object. All we are seeing is the light.

The visual point of impact occurs fifty to seventy miles above the earth, when the meteoroid reaches the mesosphere. Meeting gas molecules, compressing them ahead of the speeding mass, the resulting "ram pressure" encases the meteoroid in a glowing, superheated sheath, which usually completely vaporizes the meteoroid high above the planet's surface.

It's not friction. It's high-speed impact. The same principle used by the military to design armor-penetrating munitions.

Very few meteoroids survive their fiery plunge through the atmosphere to reach the earth. Nevertheless, it has been estimated that more than a hundred tons of material are added to the earth's atmosphere every day as a result of incoming meteoroids. That some larger, and generally more metal-reinforced, meteoroids do strike the earth is evidenced by the impact craters left in their wake.

Barringer Crater, near Winslow, Arizona, is a prime example of how big these incoming chunks of matter can be and the impact they can make. About twelve hundred meters in diameter, the crater was formed by the impact of a large (thirty- to fifty-meter class) iron meteor about fifty thousand years ago.

Impacts by larger interstellar objects, asteroids in the one-kilometer-and-above range, occur every million years or so. The damage is enough to dramatically alter climate across the planet, resulting in mass extinctions and a wholesale reshuffling of the earth's biological deck.

It is very likely that the impact of one such asteroid on the Yucatán Peninsula 65 million years ago triggered both the end of the age of the dinosaur and the rise of mammals on Earth.

Fortunately, though, few meteoroids survive their plunge through the atmosphere. If this were not the case, watching meteor showers would be considerably more than a spectator sport.

The moon was down, but the light shed by the stars, combined with the ambient glow hovering in the northern sky, was enough to guide my feet. I walk this stretch of the universe almost every morning, and my feet know every buckle and crease in the asphalt along the way.

One or two shooting stars is the usual harvest on an average morning's walk. But when the earth intersects the debris trail left in the wake of a comet, the resulting meteor shower may increase my harvest of shooting stars by a factor of ten.

For years my morning walk record was twenty-eight shooting stars. It was posted during the Geminid meteor shower, which peaks on December 13. But I've also enjoyed great mornings under the Leonids, which occur annually on November 17, and the Quadrantids on January 4, and the Aquarids on May 5.

Meteor showers are named not after the comets that spawn them but after the constellations from which they appear to emanate (in sky-watching terms, the "radiants"). The Perseids are the byproduct of a comet named Swift-Tuttle.

Every year, beginning on July 17, lasting until August 24, and peaking on August 12 or 13, the earth passes through the Swift-Tuttle debris trail. The result, while highly predictable, is magic. On average, about seventy-five shooting stars per hour can be seen given optimal viewing conditions, and conditions this morning approached this ideal.

It was three minutes before I recorded my second shooting star. Another momentary, fast-moving scratch in the sky. Classic Perseid meteors are bright white or slightly yellow, showing for less than half a second. Most are so brief, their light so frail, that they cannot be seen in urban areas, obscured by the light pollution filling the sky. Larger, brighter Perseids often leave "smoke" trails hanging in the air. It's not smoke, of course, but ionized gas in the disintegrating meteor's wake.

After several more teasers, I finally got a meteor worth ooohing about. A big, bright yellow, slow-moving one whose trajectory carried it across, not plunging into, the atmosphere. Its size and brightness increased as it shot right through Andromeda and plunged into Pegasus's rump. There is even a name for these all-star shooting stars: "Earth-grazing fireballs."

This fireball remained best in show on my first tour of the morning—one mile up, one mile back. In about forty-five minutes' walk time, I had thirty-three shooting stars for a total.

A record harvest, but I'd expected better. Commonly the greatest numbers of shooting stars are seen in the hour before dawn. It is at this time that the point of the planet we are riding on is charging head-on into the debris stream. My first walk had been just a warm-up. This one would be for keeps.

FALLING STARS, LOST IN THE GLOW
It was an hour later in the celestial rotation. Cassiopeia was almost smack overhead now, reclining seductively in her

signature W-shaped configuration. At Cumberland County's latitude, the constellation named in honor of the ancient Ethiopian queen can be seen almost year-round—low on the northern horizon in winter, high in the summer sky.

Cassiopeia is poised between Cepheus, her husband, and the hero Perseus (the slayer of Medusa and rescuer of the royal couple's chained daughter). I have always chuckled at the sanitized assertion that the constellation shows the queen seated in a chair.

Just look at those lines. And try to imagine yourself a fourteen-year-old, testosterone-charged Bedouin sheepherder lying on your back. All alone. Tending your flock. In the middle of a star-choked night. With your loins aching and your youthful imagination on fire.

Sitting in a chair. Give me a break. And I will always, secretly, wonder about the relationship between the queen and her future son-in-law. True, the Perseus of the sky, if not the one of Greek myth, is reaching for Andromeda. But he and the provocative Cassiopeia are lying on the same bed of stars.

Orion, whose three-star belt was not even peeking over the horizon when I started my walk, was almost fully visible now. Rigel, the bluish white 0-magnitude star that constitutes the hunter's right foot, was just peeking over the southeastern horizon. A giant star (on the 0–5 magnitude scale, they don't come any brighter), Rigel is thirty-three times the diameter of our sun and twenty thousand times more luminous. From five hundred light-years from Earth, its light, now passing through the portals of my eyes, started its voyage through space before Henry Hudson's *Half Moon* anchored in Delaware Bay.

If you are wondering why you are being treated to this bit of zodiac lore, I assure you it is not to impress you with how much I know. Fact is, I'm pretty much a duffer when it comes to decoding the elements of the night sky, in my estimate, barely constellation conversant.

No, the point is not to show you how much I know. It's to suggest to you how little you do. Because if you are like most twenty-first-century human inhabitants of planet Earth, you couldn't find the North Star with a star chart and Captain Jean-Luc Picard at your side.

And you're saying: "So what?"

And I'm saying, "Well, it's good you feel that way, because any chance you have to gain any intimacy with the galaxy you live in, and the lore that has enriched the lives of humans for thousands of years, is being stolen from you. In fact, over much of North America, it is just about gone. Unless you travel to a very remote corner of the globe, or live long enough for space travel to become commercially feasible, you are not going to see the skies your ancestors knew and whose elements are so much a part of our myths, legends, art, poetry, dreams . . ."

And wishes.

Of course, I'm talking about the light pollution that is eclipsing the night sky. Even if it weren't unnecessary. Even if it weren't preventable. Even if it weren't an extraordinary waste of energy. It would still constitute larceny. Your heritage is being taken from you. Your birthright as a citizen of planet Earth, stolen.

Clear, bright, star-filled skies.

The criminal thing is that you probably aren't even aware of it. And this is because this heritage theft has happened so gradually, you never knew it.

But I did and do. Saw it happening as I grew up in Northern New Jersey. See it happening in South Jersey, now.

When I was a kid in Northern New Jersey, the stars were as they are supposed to be: an every-evening matter of fact. I knew the Big Dipper. I knew the stars that formed a V (and only later found out that it was part of the constellation Taurus, the Bull). I remember the autumn night, in 1957, when the neighborhood all turned out and stood in little human constellations and looked into the sky, waiting to see the new star that had been shot into the heavens. A star that moved in orbit around the earth. A star named *Sputnik*.

And how hard it was to see that tiny, man-made star against the backdrop of God-made stars. If it hadn't been moving, you'd never have picked it out at all.

I was six. Route 10 was a two-lane highway, and it cut through woodlands. Parsippany was farmland. Morris County was not the corporate headquarters capital of North America. The glow in the eastern sky, which was there all night like some premature dawn, well, that was "over there."

If I looked at a map of New Jersey, I could see "over

there" defined. It comprised the counties of northeastern New Jersey—Essex and Bergen and Hudson (vassals of New York City). It was a land of streetlights and traffic and cities and people. On the map it was shown all in yellow.

But where I lived, the map was all in green.

I've always loved green. And I think it was precisely because urban areas on maps are tinted with yellow that I developed my antipathy for the color. When I first read J.R.R. Tolkien's Lord of the Rings, and got to the part where Faramir is talking to Frodo and he nods toward the east, toward the land of Mordor, and says, "Yon" (not even wishing to honor the evil place with a name), well, I forevermore thought of that blighted land east of Whippany as Yon.

My disdain for Yon was not linked specifically to light pollution. I thought in terms of natural areas and unnatural ones, rural areas versus urban. What I didn't appreciate was that I was growing up in a battleground. A forming landscape that was not urban or rural but something in between. It was called "suburbia."

This postwar no man's land was being invaded by people who craved rural space but demanded urban amenities.

By large corporations and distribution centers that wanted to get away from the congestion and taxes of the cities. And they brought outside lighting to illuminate their parking lots.

By people moving into the suburbs (because that was where the employment went). And they bought houses with lighted patios (so they could enjoy that open space after work) on streets named Sky Line Drive and North Star Way.

The streets, of course, needed lights so that people could feel safe. So did the shopping centers and malls that serviced these new suburbanites and the schools that had to be built to educate their kids.

Lights intended to illuminate the earth, but whose light radiated, by reflection and poor design, back into the sky. Where it didn't just keep going, out into space. No, it became trapped and redirected, "scattered" by moisture and dust so that it hung in the air like a milky shroud. On cloudy nights, it was even worse. The light would be reflected by the clouds and bounced back at the earth.

You could read a newspaper by this boomeranging light. You could kiss good night to night.

I didn't realize, then, what all this light was doing to the stars I'd known as a kid. It happened too gradually. In 1977 I moved to South Jersey, where the stars were still bright, and I discovered I'd been robbed.

From the vantage of mounting years, I looked at maps and saw the yellow smear slowly creep across North Jersey and cover the places I knew. I drove home to visit my parents along four-lane highways that used to be two. At every visit, it seemed I stopped at new traffic lights erected at old intersections and past sprawling strip malls that used to be supermarkets. I came to realize that by the 1980s my childhood home had become Yon.

Oh well. I lived in South Jersey now, where the stars were still bright (particularly when all the summer people packed up, turned off their patio lights, and went home). The people who lived in South Jersey were also clearly smarter than the people in North Jersey. They valued their open space. They didn't want their freedom and mobility hampered by rampant growth and development.

They used to tell me, with adamancy and pride, "Well, we're not North Jersey." And I believed them. Believed that they could see what had happened in the northern part of the state (as plainly as I had) and were committed to not letting it happen here.

In 1990, Linda and I moved to Mauricetown. I started my morning walk routine soon thereafter. My one-hour walk at Turkey Point, beginning under the stars, ending near dawn. And I was delighted, but not surprised, to see, as an every-morning matter of fact, the stars I'd known as a child.

But little by little, by incremental encroachment, the same thing that had happened in North Jersey began to happen here in South Jersey.

Despite the example. Despite the presumed wisdom of South Jersey residents. Despite the "we're not North Jersey" resolve.

Cape May County's population burgeoned with retirees. Now, on cloudy nights, you can stand on Turkey Point and see the peninsula outlined as a milky white projection in the sky. Expanding prison populations demanded more facilities to hold them, and now, to the east of Turkey Point, the greatest blemish of light on the horizon is the one built to hold criminals, while its poorly directed light robs the heavens blind.

The sand-mining industry opened new pits, with batteries of lights so that they could operate at night. And while the lights projected down, onto the pits, the water in those pits acted like a mirror, blowing the light back into the sky.

And nine miles north, in Millville and Vineland, the malls and shopping centers and car dealerships and their night-canceling batteries of lights followed the opening of Route 55 South (just as suburbanites, in North Jersey, followed the interstates west). Now, the skies lying north of Turkey Point are death to stars.

In less than twenty years, seeing the Milky Way has gone from being a matter of fact to being a matter of conversation in household Dunne. The pollution of the night sky has been that fast, and that complete.

Those proud old residents who told me "Well, we're not North Jersey" were right. In North Jersey, people could honestly say they didn't realize what happened, didn't appreciate the price being levied against their birthright. People in South Jersey don't have the luxury of this excuse. They had the example right there in front of their eyes. Knew that the price of growth was their heritage and that part of this dowry was the diamonds of the night sky. Like their North Jersey kin, they traded diamonds for tarnished silver. Their heirs will now grow up as ignorant of the stars and as shooting star–deprived as the children of Yon.

And for no reason! Ways of addressing and redressing the problem have been and continue to be developed. There are, now, outside lighting fixtures that dramatically curtail the amount of light lost into the sky—saving energy, saving money. Hundreds of communities have come to realize that light pollution is as much a threat to people's quality of life as water, air, and noise pollution. There are, now, statewide and municipal laws requiring outdoor lighting designed specifically to reduce light pollution.

It's not just an aesthetic appreciation of the night sky that is at stake. Light pollution intrudes into the privacy of homes. It misdirects and causes the deaths of tens of thousands of migrating birds. It disorients nocturnal pollinating insects (like moths). It wastes millions of dollars.

All that light, going into space, serves for nothing. More concentrated and better directed light greatly reduces the energy needed to light a targeted area. This means you need fewer lights and lower wattage to achieve the same results.

But even if the only sad consequence of light pollution was erasing the stars from the sky, wouldn't that be enough reason to stop it? Have you ever considered how much stars have meant to our species? How much they mean to you?

We elevate the people we most admire to their status. We call them "stars."

When you turned in a particularly impressive homework paper, what did the teacher stick on the top that made you so happy? Most of the time it was a star. Sometimes the American flag. But what do you find in the upper-left-hand corner of that flag?

Exactly right. Fifty of them. One for each state.

When soldiers are recognized for their valor, they receive a bronze or silver star. For conspicuous gallantry they receive the Congressional Medal of Honor. Guess what's on it?

We name sports teams (like the North Stars) after them.

We sell products with them. Did you know that the company logo for Subaru is a star cluster known as the Pleiades, which is found in the constellation Taurus?

What was it that shepherds saw hanging over the town of Bethlehem?

What do you hitch your wagon to when your ambitions are soaring?

What do you wish on when you want something very much?

And at the beginning of my walk, when I stepped out of my car, you'll remember I made a wish. Unlike your birthday wish on candles, I don't think that there is any clause associated with starstruck wishes that renders them null and void by divulgence. In fact, I'm more afraid that keeping my wish to myself will render it void.

So I'm going to divulge my wish. I'm going to share it with you.

What I wished was that the people in Cumberland County, and the representatives they elect, and all the people in all the places where natural heritage is prized, would make the protection of their skies a priority. That municipalities would adopt legislation to corral outside lighting. That businesses and corporations would take it upon themselves to reduce the amount of light pollution their enterprises create and restore to people their heritage.

I saw forty-three shooting stars on my second walk. A new record for Turkey Point. Included in their ranks was one buster of a fireball that I didn't actually witness. It exploded behind me while my back was turned. But the strobe flash of its explosion lit the road and marsh around me.

When I turned, all I could see was a cloud of smoke hanging in the air.

Maybe I'll catch the next one. Then again, unless a few ordinances are enacted sooner, my record of forty-three shooting stars at Turkey Point might stand a long, long time.

Until the next once-in-every-65-million-year asteroid arrives. And turns out the lights.

CHAPTER 10
Protectors of the King's Deer

After deliberating, I chose the twenty-four-ounce coffee over the twelve-ounce. It was going to be a long night. In fact, I didn't have any idea when we'd be getting back—and I told Linda as much.

It was a Friday evening in late August. The Wawa, crowded, as usual. I paid for the coffee. Got to my car without scalding myself. Set the coffee on the roof. Started rifling through my daypack to make sure I'd included a writing pad. Turned to find Greg Honachefsky, seated in a black SUV, waiting for me to notice him.

I never heard him drive up. And I wondered how many poachers, over the course of Greg's twenty-year career in law enforcement, have looked up, seen the New Jersey State conservation officer looking at them, and wondered where the hell he'd come from.

"Am I late?" I asked, searching for space in the four-by-four's crowded interior, doing my best not to dislodge the twelve-gauge Remington riot gun poised between the driver and passenger seats.

"That's new since you came out with me last," he said, reading my mind, referring to the last stakeout I'd done with the then junior game warden nearly a dozen years ago. Now a lieutenant, he still spends most of his life away from the desk and in the field.

Greg's job is catching bad guys. The kind that take liberty with the natural resources of the state. Your resources. His authority, like his job, goes all the way back to a time when America owed fealty to a king, and his vocational ancestors used to catch slayers of the king's deer with leg traps.

Nowadays, poachers are merely fined, jailed, and relieved of their legal hunting privileges.

We turned left out of the Wawa and toward Mauricetown. I didn't know where we were headed, except that the site was "a tree farm near Greenwich" and that, based on our trajectory, we were going to be driving right past my house.

"Hell," I said, "if you knew we were going this way, you could have just picked me up at home."

"I don't know where you live," Greg said mildly, and this disclosure is probably a good thing. If the game warden knows where you live, it's probably for the same reason you wish he didn't.

Game wardens do not enjoy this kind of anonymity themselves. Forty-eight hours after a game warden moves into a bayshore community, everyone knows who he is, where he lives, what kind of vehicle he drives, and pretty much where he is at any given moment.

It's not that South Jersey is lawless! It's more nearly correct to say that old-time residents are naturally cautious. People who have, for many generations, taken much of their livelihood from the land find themselves, at times and for any number of reasons, at odds with authority that blunts their effectiveness.

"Have you ever busted a woman poaching deer?" I asked.

"Nope. Never."

"How come?"

"Testosterone," Greg assessed.

As we crossed the Maurice River bridge, Greg got on his radio and contacted his partner, advising him that we were "twenty minutes out."

"Who's 'we'?" Zane wanted to know.

"Pete Dunne," Greg said.

"Who the hell is Pete Dunne?"

Apparently my anonymity among game wardens was broad-based. More good news.

NORTH JERSEY TRANSPLANT

Greg Honachefsky, age fifty-two, was, like me, a North Jersey transplant. Born into a hunting family in rural Hunterdon County, New Jersey. Six foot one; 195 very solid pounds, he looks like Paul Newman and moves with the stiff, precise grace of a platform diver. He is, like all New Jersey conservation officers, a graduate of the New Jersey State Police Academy. He was, for many years, defensive tactics instructor for the New Jersey game wardens. He also competes, annually, in the Fourth of July race up Mount Marathon in Seward, Alaska.

He is also an avid reader, plays the violin, volunteers for the Bayshore Discovery Project, performs in a Celtic band, and stopped hunting long ago.

Hunting animals, anyway. When his life became hunting people, hunting animals became something of a pantomime.

Zane Batten, Greg's partner, is different. A serious hunter and a big man, well over six feet, pushing 240 pounds, Zane was dressed all in camo, standing in the driveway of his home and leaning against the side of a vehicle identical to Greg's when we drove up.

We shook hands, and my anonymity slipped away. Even in the deepening twilight I could see Zane studying my face. Facts and faces are part of a conservation officer's stagecraft. Add to this patience, perseverance, intelligence, intuition, and a stable of informants, and you have the makings of a successful career in law enforcement.

But there are also times when being six foot four and 240 pounds is a real vocational asset.

"I found a place that gives us a little high ground

tonight," Zane said. "We're away from traffic and houses. We can back up between the rows of the orchards if we see lights," he said.

Just like in hunting, most of the effort that goes into a successful stakeout is on the front end. If you know your quarry, if you know the territory, and if you have a good plan and anticipate contingencies, things have a way of going right more often than not.

"I'm feelin' lucky tonight," Greg said, after hearing Zane out.

"Yeah, me, too," said Zane.

That's like hunting, too.

The interesting thing, maybe the fascinating thing, is that both Greg and Zane knew who they were hunting; knew, in fact, just about everything about their "client."

Knew where he lived. Knew what kind of rifle he fired. Knew his MO—when he hunted, where, how he located and shot the deer he took illegally, which direction he'd be coming from, whether the bust would be orderly or there was going to be a chase.

The guy they were after tonight was a well-to-do businessman from central New Jersey. He knew the people who owned the orchard and took covering advantage of the deer depredation permit that the owners were issued seasonally. Used it to take deer—bucks—illegally.

Killing the deer that were using the ornamental trees to "rub"—that is, mark their territories by gouging divots in the trunks of trees—was not illegal. The farmers were issued permits to combat the economic loss caused by deer. What was illegal was shooting the deer with a spotlight, at night, with a rifle, without being one of the people named on the permit.

Any one of those infractions is a bust-worthy offense.

"How'd you find out about this guy?" I asked, wondering whether it was something I should ask.

Zane looked at Greg, letting him decide what to say or not to say.

"Informants," Greg said, answering broadly. "We depend a lot on informants."

So, I thought as we retreated to our vehicle. The relationship between Greg and Zane and their "clients" was not simply predator and prey. It was cat and mouse and a complicit rat.

Zane went off to pick up the last member of the team, a young, recent graduate of the conservation officer course named Keith Fox. He would be serving in the capacity of apprentice, and this was his first stakeout.

Half an hour later, under a moonless sky, we rendezvoused again and drove to a two-track road skirting the edge of the orchard. Backing between the rows of trees, we got out, made brief introductions, and got a quick lay of the land.

Zane stepped onto the hood of his car, which protested audibly. Greg slipped a stick of nicotine gum into his mouth. We settled into the time-honored process of waiting. It's what hunters (and wardens and poachers) do most.

"Why is that telephone pole lit up?" Greg said, coming to a level of alertness that defined the difference between alert and riveted.

At the far end of the row of trees, a light was shining on a utility pole that moments before had been dark.

TIME-HONORED TRADITION

Poaching, the act of catching or killing game illegally, is a time-honored tradition on the bayshore. As an institution, it goes back to a time when all deer were considered the property of the king, and killing the king's deer was a capital offense.

The risks were high. But people still did it. In fact, some, like Robin Hood, are still celebrated for their audacity and snub of authority. When Europeans first settled in the New World, they discovered a wealth of living, harvestable things. Deer and elk, fish and fowl. An "unlimited" bounty, there for the taking. To land- and resource-starved Europeans, it seemed too good to be true, and it was. As the myth of North America's unlimited natural bounty began to tarnish, the king's law was reincarnated as state law, so that, no matter who owns the land, wildlife is the property of the state. As early as the 1700s, New Jersey had laws on the books regulating the manner in which deer and other game animals could be taken. Even so, by the late 1800s, the laws were mostly superfluous. There were hardly any deer worth poaching.

In fact, at the beginning of the twentieth century, in all of New Jersey, there were fewer than fifty "wild" deer.

This disclosure will come as a shock to today's New Jersey suburbanites, who wage a ferocious war against shrubbery-munching white-tailed deer, and whose disproportionately high auto insurance rates directly reflect the high cost of highway collisions with deer. But deer were not the only animals whose populations had been nearly depleted as a result of overharvesting and habitat loss. By the end of the nineteenth century, shorebirds were a vestige of their former abundance, too, their migratory populations shattered by the guns of market hunters.

Waterfowl numbers were hardly better. Turkeys were extirpated. The last black bear killed in New Jersey was a matter of public record. "Sportsmen" were reduced to killing songbirds like robins, flickers, and bobolinks largely because these were the only targets whose size and numbers made them marginally worth shooting.

It took a national outcry, a whole "conservation movement," to put America's wildlife back on track. Many of our game laws, many of our state and federal agencies, and not a few conservation organizations (including the Audubon Society, the Sierra Club, and the Izaak Walton League) have their grounding in the wildlife dark ages of the late 1800s.

Through active protection, targeted habitat manipulation called "wildlife management," and, in the case of deer and New Jersey, reintroduction efforts, involving deer brought from states whose populations remained more viable, New Jersey's deer population, and North America's shorebird and waterfowl populations, gradually increased.

So did poaching. In fact, long before New Jersey saw fit to open a hunting season on deer, venison was back on the menu in many rural parts of the state. In South Jersey it became a cultural institution. A tradition handed down father to son; a twentieth-century recasting of Robin Hood and his Merry king-snubbing Men.

Tonight, in this South Jersey orchard, I was standing next to the Sheriff of Nottingham. Somewhere in the woodlands were the outlaws—the slayers of the state's deer.

IN MERRY OL' NEW JERSEY

The illumination faded. The car, whose headlights were shining on the pole, turned them off or turned them away. It might have been someone just turning the car around, or it might have been a poacher whose intuition or luck was good and who took his king-snubbing ambitions elsewhere. In any event, the source of the light was never revealed, and slowly the four of us relaxed, falling, in the social pattern of our species, into quiet conversation—most of it centered on celebrated deer busts of the past.

Greg and Zane took turns, offering accounts of high-stakes (and often humorous) engagements with the bad guys.

The time, on a tip, they went up to a barn while a merry band was busy butchering deer killed out of season. Knocked on the door. Were invited in.

There, caught very literally "red-handed," were six very surprised outlaws along with eight illegally taken deer.

Nobody even tried to run. And it's tough to deny culpability when you are standing in a barn with a haunch of state-owned deer in one hand and a hacksaw in the other.

Or the time Greg staked out an enclosed deer stand the day before opening day. He watched as preparations were made—blind stocked, bait laid—and noted that, while two men drove in, only one drove away. Climbing the ladder into the stand, after dark, he came face-to-face with a very contrite gentleman with a spotlight and a loaded rifle, who was probably quietly cursing the diligence of wardens.

Killing deer over bait is legal in New Jersey. Killing deer before the season, after legal hours, using a spotlight and an illegal weapon is not.

The one I liked best was the guy who butchered an illegally killed deer, stuffed the hide and scraps in a box, and dumped it at the end of a dirt road. A box from Cabela's, a large mail-order chain specializing in hunting equipment, with an address label that led, it turned out, not to the man who killed the deer but to the man who helped him butcher it.

Who, then, obligingly turned the shooter in.

Most poachers are not this dumb, and some are exceptionally cagey. One man, whom Greg pursued for years, poached for a living. He'd take orders in advance and hunt only in places he knew intimately, far from the road. Baited and killed deer while walking a route, at night. Butchered them on the spot. Packed the meat in bags, which he would then deposit just off a county road in proximity to a numbered telephone pole whose location he divulged to clients over the phone.

In spy versus spy parlance, this is called "a drop."

But the guy we were after tonight wasn't interested in meat. He was interested in racks, or antlers, and, according to Zane, fifty bucks were taken "on the farm permit" last year. This guy was in it mostly for the bragging rights killing a deer with a big rack confers.

"Testosterone," as Greg had said.

Translation. Mine's bigger than yours. And when they're hanging on the wall, racks taken illegally look just the same as those that are taken legitimately.

SEEDS OF AN INSTITUTION

The moon came up around 9:30. A bright moon. Almost, but not quite, good enough to shoot by without a spotlight. The only sounds came from the trilling chorus of crickets and the high-pitched clicking of bats. Except for the adrenaline burst triggered by the illuminated pole, nothing had happened to stir the tranquillity of the night.

Greg and Zane began to speculate on what was, or more specifically what wasn't, happening. It was a stakeout. Nothing more. There was no specific intelligence to suggest that the poacher would be active tonight.

"How many stakeouts result in busts?" I asked.

"One in eight," Greg said, more quickly than contemplation could account for. "I calculated it once," he explained.

This means many hours of waiting and frustration for a couple of hours of success and excitement. Poachers enjoy a much higher success rate, and it explains, in part, why killing deer illegally is widespread enough to command a force of sixty state game wardens to keep it in check.

Greg musingly noted that one day, not long ago, he was driving through a community in his jurisdiction and grew increasingly aware that he'd met, in his professional capacity, the residents of just about every home he passed. "Did a search warrant on that one . . . staked out that guy's place for weeks . . . caught that one with two illegal does . . . that one trapping otter illegally . . ."

In deference and fairness, it should be noted that there are hundreds of state and federal game regulations. Regulations that cover everything from bag limits to shot size and type to how much orange a hunter is required to wear to how a deer must be identified and tagged before it can be moved.

A hunter who goes hunting for rabbits in the morning and ducks in the afternoon but fails to remove all the lead shot shells from his pockets (only nonlead shot may be used to hunt waterfowl) is in violation of the law. A deer hunter who gets into his tree stand an hour before sunrise and loads his gun as soon as he is settled so that the sound and motion of loading up will not alert deer when legal shooting time arrives is in violation of the law.

But these infractions, while violations of the game code, are misdemeanors. Poaching deer, at night, is a crime at a higher level of magnitude. I've never done it. I've never even wanted to do it. But I have wondered about its appeal or, put another way, how is it that poaching came to be such an institution of the bayshore?

First off, and maybe needless to say, poaching is akin to hunting, and hunting is a time-honored tradition across all of rural North America. It is a way people have engaged their environment for thousands of years. It involves nature. It involves father-to-son family tradition. It strums atavistic chords in our genes and in our blood that lie deeper than understanding.

There are lots of people who hunt. In New Jersey alone, there are ninety thousand licensed hunters. While not all hunters are poachers, it is fair to say that all poachers are hunters. Brigands, too. But hunters.

Fundamental to all human endeavor is the desire to succeed, and the problem with game laws is that they are designed to protect wildlife populations by blunting human success. They do this by prescribing the sex, age, and number of deer a hunter can kill, as well as not allowing hunters to kill deer at night, when they are most active.

For many years, in New Jersey, a hunter could take only one antlered deer a season. The season ran six days in the first week of December. Legal hours were 7:00 A.M. to 5:00 P.M. If you didn't get a buck, tough. If you got a buck but wanted to take a bigger buck, you had to break the law.

It was very, very limiting, and it flew in the face of a lot of human ambition.

It was also very successful. Deer populations grew. In rural South Jersey, with its hunting tradition and its economic stagnation, it was too much to hope that such an attractive resource would be overlooked.

Because the area was large and unpopulated, because

deer were killed largely at night, because wardens were comparatively few (compared with jacklighting hunters), the chances of getting caught were low. Before Greg came on the scene, the situation in South Jersey was pretty wild and woolly. Older game wardens spoke casually of running gun battles with poachers. When Greg moved into Dorchester, his welcome letter was spray-painted on the street in front of his house.

It read: $5000 REWARD FOR SHOOTING THE NEW WARDEN.

The problem Greg and other wardens faced was not one of identifying poachers. Through undercover work, through informants, Greg and company pretty much knew who was poaching deer. As Greg said, often he'd chase a person all night, then find himself chatting with him at the Wawa in the morning.

He recalled another surreal encounter, in an old cinderblock bar that used to operate just south of Bivalve. It was New Year's Eve. The poachers weren't poaching. The conservation officers weren't busting. In the spirit of "Auld Lang Syne," his "clients" were buying him drinks.

While I was writing this chapter, during a routine visit to my physician, a lifelong Vineland resident, I mentioned my interest in poaching and was surprised when the doctor allowed that two of his buddies in high school were then, and for years after, accomplished poachers. One boasted that he'd never killed a legal deer in his life and that he'd never been caught.

And I think here, in this boast, is a bit of insight that sheds light upon the institution of poaching. Yes, certainly some people poach for food, but added to this, and maybe not subordinate to this, is the fact that poaching is challenging and exciting—exciting enough that it confers bragging rights.

If hunting is fun, how much more exciting is being hunted? Outwitting the agents of the state who are imposing their unpopular laws. The Sheriffs of Nottingham. Or, as another bayman, a resident of Leesburg, once reminisced to me about his boyhood of truancy along the banks of the Maurice River, "the best-tasting watermelon is a snitched melon."

"Did you ever find out who put a bounty on you?" I asked Greg.

"No," he said. "But so far nobody's collected."

The Death of Poaching

The Pleiades star cluster was just peeking over the eastern sky when the radio crackled and Greg took a call from Doug Ely, a conservation officer whose beat covers Cape May. Doug, too, was on a stakeout. Doug, too, was not having any luck catching bad guys this evening.

Greg looked at his watch, which confirmed what the position of the stars was proclaiming.

"Eleven," he said for the benefit of those who hadn't looked at either their watches or the sky. "What do you think?"

"Go to twelve," Zane said.

Greg was senior officer. But it was Zane's play. We'd stay till midnight.

Poachers, like deer, are largely crepuscular—most active

just after sunset and again before sunrise. They might stay up late dealing with a deer they've killed, but most illegal shooting occurs earlier in the evening, not later. It was looking like this night's effort would fall among the majority—a night without a bust.

For any number of reasons, these bustless nights are becoming more common, even though Zane estimates, and Greg agrees, about 30 percent of the deer taken in the state of New Jersey are taken illegally.

The reason is very simple. There are fewer active poachers than there used to be. Like so many other bayshore traditions, the art of poaching is a dying institution.

Even though the economy is faltering. Even though unemployment is increasing in Cumberland County. Even though testosterone is no less determining and egos no less driven than they were in Granddad's time. Even though deer are more common now than ever, and in places their population has reached the level of a plague.

In fact, the current abundance of deer is directly related to why poaching is on the wane. In this age of ungulate abundance, there is not only an abundance of deer but also nearly unlimited opportunity to hunt them. Today a hunter in New Jersey has, through various seasons and licenses, almost six months of deer-hunting opportunity.

Hunters are still limited in the number of bucks they can harvest. In some places, though, the limit on does is zero—meaning no limit! In fact, in some "zones," hunters can shoot two deer at a time, check them in, and go out and shoot two more.

Closely related to this abundance, the sheer volume of deer being taken means that it is easier to fudge. Poach by omission, not by commission: Fail to tag your small buck, and drive it home without stopping at a check station (so you don't waste a valuable buck tag). Take a deer in one zone and report it in another, whose season is more liberal or extensive.

The people who are shooting deer, at night, with spotlights, are the guys who are driven to shoot deer with impressive racks. Really big deer get that way by being smart, and part of being smart in a heavily hunted state means being nocturnal. In addition, and quite unlike those who hunt by the book, poachers get a jump on seasons, and, of course, people who pay little attention to legal hours or legal dates are not likely to pay attention to some niggling little details like "bag limits."

The guys who are in it for the antlers kill multiple bucks. It is these rack-driven poachers who constitute most of Greg and Zane's "clients."

That testosterone thing again.

Two other factors have contributed to the decline of poachers—neither of which was mentioned by Greg or Zane. The first is their success. In their tenure, eighteen years for Greg, nine for Zane, they have busted scores of poachers and harassed others into retirement or retreat. One former "client" moved to North Dakota to get out from under Greg's eye.

The other reason poaching is on the wane is that poaching, real poaching, is hard work. You are going to get clawed by catbrier. You are going to be infested with chiggers. You are going to drag dead deer in the dark. You are going to have to stay up late, cutting deer into edible (and concealable) packets.

I stopped by the Mauricetown Gun Club one morning. Caught up with a member. Asked him to explain the decline in poaching (making it clear I was not implying that their club or its members were engaged in any illegal activity). He mentioned the undermining influence of liberal seasons, and the success of conservation officers.

"People don't poach because they know they are going to get caught. There used to be a few guys doggin'," he added (running deer with dogs as if they were just running rabbits), "but they don't do that no more."

Then he mentioned his son. Sixteen years old. Goes hunting with him, "some." But the boy was so overweight that he couldn't "move easily through the woods."

"When I was his age," the club member continued, after a long and thoughtful pause, "I couldn't wait to go hunting. After school. Weekends. But he's more into his computer games."

I guess I should be happy that poaching deer is a dying institution. And I guess I am. But I'm also disturbed.

That killing people and stealing cars on a computer is preferable to killing deer for real. Not because it's more socially acceptable but because it's effort-free.

That a whole generation may grow up never even wondering, much less discovering, whether a snitched melon tastes better than a store-bought melon.

Orion was lurking just below the horizon when we called it a night. Said our goodbyes. Got in our respective vehicles and headed home. We drove with our lights off until we got on pavement, but the moon was up and the two-track easy to see now.

We drove in silence for a time, both of us tired and contemplative. It was Greg who broke the silence, concerned, perhaps, that a night without action would undercut the story.

"We don't get everyone, on our own terms, or everyone every night. We plan to, but we eventually do get everyone. There isn't a man who can't be caught with a wee bit of effort, persistence."

I nodded. I hadn't picked the wrong partners, just the wrong night.

The Wawa parking lot was nearly empty of cars when Greg dropped me off. Late-shift change time at the prison had come and gone. We were entering the period when human activity, like deer activity, is at its nadir.

It was time for wardens, poachers, and writers to go to sleep.

Tomorrow night was another day.

CHAPTER 11
Tramping with Dallas

By the time I reached Haleyville, the clouds spawned of Hurricane Fay, the fifth named tropical storm of the season, had gained full control of the sky. Rain was not far off.

Located astride the Haleyville–Dividing Creek Road, scarcely a mile west of Mauricetown, this cluster of nine-teenth-century homes lies mostly within the shadow and ecumenical orbit of the Haleyville Methodist Church and its welcoming cemetery. As if to underscore this, a man driving a winch-bearing truck was parked in the northeast corner of the cemetery, confirming that a new resident was about to be welcomed into the community this very morning.

Haleyville's living community is comparatively modest: fewer than thirty homes and perhaps one hundred living souls. The body of those interred in the community's cemetery, however, is large—numbers over a thousand—and to look at their names etched in limestone, marble, granite, and bronze is to read through a veritable who's who of the Delaware Bayshore.

Scattered among the stones are those bearing the name Hand, representatives of a family whose lines go all the way back to those pioneering Town Bank whalers of the 1600s, and here and there are Petersons, residents whose lineage can be traced to those pioneering Swedes who settled first in Salem County and later on the banks of the Maurice River.

There were Robinses from the hamlet of Robinstown and Fergusons and Comptons and, as was to be expected, an impressive number of Haleys. But ranked among those earliest families is a passel of Lores and Sharps. It was one of these Sharps that I was coming to see: Dallas Lore Sharp. I had it on good authority that this was his current address.

The Sharps of Haleyville

It was Hezikah Lore who came to this section of land in 1722, establishing the gristmill that became the economic focal point for the town of Underwood, later renamed Haleyville. The store that served the community was operated by R. L. Sharp.

As evidenced by the mill, Haleyville was a farming community. But because it was just a mile from the shipbuilding center of Mauricetown, not a few local lads were drawn into the alluring orbit of that place and to lives at sea. If you look at the older monuments in the graveyard, distinguished as much by their close proximity to the church as by the rich crusts of lichen that encase them, you'll find a number of Sharps and note that many are honored with the title Captain.

This evidence, and the short span of years most of these early Sharps enjoyed on this earth, suggests that a number of R.L.'s ancestors did not enjoy the long, prosperous lives of shopkeepers but were, instead, drawn to lives of adventure. The Sharp I was searching for this cloudy morning at summer's end fell into this adventurous lot.

Dallas Lore Sharp. Born December 13, 1870, in Haleyville to Mary Denn Bradway Sharp and Reuben Lore Sharp. Considered by the famed nineteenth-century naturalist John Burroughs the best nature book writer of their age, the future biologist, minister, college professor, and author of over a dozen books ranged all over the woodlands, farmland, and, especially, the marshes that were within walking distance of his Haleyville home. His favorite haunts, as evidenced by his writings, included East Point, Bear Swamp, and the Maurice River.

Places that Linda and I love, too.

He did, in the course of his life, travel to England and Panama, and, near the end of his days, he lived for a short time in Santa Barbara, California. Most of his adult life was spent not in Cumberland County but on a small farm in Hingham, Massachusetts, not far from Boston College, where he was an instructor of English.

But Cumberland County was where Dallas Lore Sharp acquired the bond with the natural world that was the foundation of his written works. In his book *Sanctuary! Sanctuary!* Sharp observed that "no boy can pick his parents or the place in which he ought to be born . . . but if I could have chosen, I should have picked the very parents I had and the same unheard of place where I was born."

Never heard of him, you are thinking. John Muir, yes. Burroughs, the Seer of Slabsides? Absolutely. Sharp? Uh-uh.

Well, it might interest you to know that *The Seer of Slabsides* was the title of the book written by Sharp about his older friend. And unless you are a student of the conservation movement, that great turn-of-the-century shift in environmental thinking spearheaded by people like Teddy Roosevelt, John Burroughs, and John Muir, you may be forgiven for never having heard the name of their esteemed contemporary.

I hadn't. Not really. Not when Linda and I began this project. Not until I started to speak with some of the older residents of the region, who extolled Sharp's fame and virtues.

So it was only natural that I decided to learn a bit more about the life of this Cumberland County favorite son, whose affinity for the natural riches of the region equaled my own. What I discovered was not only surprising. It was eerie.

Not only did Dallas Lore Sharp write extensively about nature, but he even wrote a four-book series based on the seasons! *The Fall of the Year, Winter, The Spring of the Year, Summer.*

And all this time I'd been afraid that Linda's and my effort was going to be considered an echo of Edwin Way

Teale's *North with the Spring, Journey into Summer, Autumn Across America,* and *Wandering Through Winter.* The kid from Haleyville had beaten our effort by a century and Teale by fifty years!

It gets even better. Sharp and I share the same publisher, Houghton Mifflin Harcourt (Teale, at least, was hooked up with a rival house).

But I think the thing about this now obscure teacher and naturalist that astonished me most and endeared me most was his very clear and obvious focus on getting people—most specifically young people—out into the natural world. In fact, his four books on the seasons were printed in "school book editions" with instructional notes and suggestions for classroom use.

The final book in the series, *Summer,* is both a literary and a literal blueprint for getting young people out and engaging the natural world. At first I felt pretty chagrined that what I thought was such a novel and timely idea was, in fact, literary leftovers. Then I began to wonder why, if the natural wonders of Cumberland County had been extolled more than a century before, so many people (both young and old) were still driving right by one of the greatest natural places on the planet.

Then I began regretting never having had the opportunity to talk with Sharp, one writer of natural history to another, about our work and the subject that troubled us both—humanity's growing estrangement from the natural world.

And then, while pawing through the archives of the

Mauricetown Historical Society, I ran across the article written about Dr. Sharp by Verna Bosshart in the *Bridgeton Evening News* in 1964. It was a very good article, a fine treatment, and in it I found the author's current address.

It was the Haleyville Cemetery. After his death in 1929, the naturalist's ashes were returned to New Jersey and interred near his boyhood haunts. My uncelebrated literary twin resides a mile from our home.

I drive right by his place every morning!

Looking Sharp

I parked in the middle of one of the narrow lanes cutting through the cemetery. The place was clearly designed for residents and pedestrian traffic. I was glad my Subaru Outback's wheelbase wasn't wider.

Dr. Sharp's residence, as given in the article, wasn't down-to-the-stone precise, so I began running through the lineup of markers, searching for Sharps. Many Sharps later, I still hadn't found the one I was looking for, and, in retrospect, my problem was understandable. What I was searching for was a monument whose stature was equal to my esteem. What I found after an hour's search was a marker more in keeping with a very modest man. A small brass plate, set in the ground, commemorating and tersely summarizing the naturalist's life.

DALLAS LORE SHARP
Born December 13, 1870, Died November 29, 1929
Naturalist, Teacher, Author

"Dr. Sharp," I said to the metal casting. "Mind if I intrude on your privacy?"

I don't know about you, but I've always been a little uncomfortable initiating conversations with the dead. It's not the being dead part that troubles me. It's just that I'm never quite sure to what extent my concerns relate to their concerns. I mean, classic conversation starters, such as "How are you doing?" or "Have you got a minute?" probably don't have much meaning to people who have little interest in time and whose fortunes don't change much day to day.

I tried and succeeded in calling up the image of the man who was, in life, Dallas Sharp. A photo of a boyish but studious-looking young man with a full mane of hair, parted in the middle; dark, intelligent eyes set behind small wire-rimmed glasses, a sharp nose, full, fleshy mouth. He liked to wear full bow ties when he lectured. In later years, as his hair thinned, he took to wearing a short-brimmed cap. In a suit, he might easily have been mistaken for a banker or an undertaker.

"My name's Pete Dunne," I said. "Neighbor. Live right behind the Methodist church in Mauricetown."

This much was true. Sharp and I have even our residential proximity to churches in common—right down to the denomination and minister (the Mauricetown and Haleyville churches share a minister).

"That's the new church," I added. "The one built in 1880." When Dallas was eight, his mother and new stepfather moved to Bridgeton. The new church was built after he moved. The old one burned to the ground in 1932, having served out its last years as a pool hall.

The Mauricetown that Dallas knew, and whose shipyards served as his playground, was, then, the largest town in Commercial Township. Haleyville, which boasted a railway station that was called, to the consternation of Haleyville residents, the Mauricetown Station, constituted the suburbs. The river, where Dallas spent so much time, was, therefore, about two miles from home. The marshes of East Point, which Dallas greatly loved and wrote about in his book *Roof and Meadow*, were a twelve-mile walk (one way).

If you were wondering why a town the size of Mauricetown required two shoe repair shops, here is part of your answer. People did a lot more walking in the days before the internal combustion engine came to dominate transportation. If the Wawa had been around in 1875, Mauricetown and Haleyville patrons would have walked to buy their lottery tickets.

There were seven hundred residents in Mauricetown during Dallas's years, the town's heyday. By 1932, it was down to three hundred—about the same number that were there when Linda and I moved into the town.

All living. The Mauricetown Methodist church has no graveyard. When residents of Mauricetown die, they move to the Haleyville suburbs.

"Listen," I said to the essence of the man beneath the plaque, "I'm a fan of your writing and a colleague.

"Make that a competitor," I added. Why lie to the dead?

Neither my overture nor my admission made much of

an impression upon the naturalist, teacher, and author. But people who engage the natural world tend to be somewhat patient and contemplative, so I concluded that the august naturalist, by his silence, was inviting additional information.

The alternative explanation was that he wasn't listening, but as anyone who has ever been misquoted by a writer will tell you, that is a very dangerous assumption to make.

"Anyway, Mr. Sharp, my wife and I are working on this book project that's intended to showcase all the really great natural history of the region with an eye toward getting people jazzed about going out and seeing what they are missing.

"*Jazzed,* by the way, means 'excited.'

"We didn't realize that you'd plowed this ground already. Written so eloquently about nature and how engaging it is, particularly for young minds. But it could be argued that it's a good thing Linda and I came along because . . . well, you see . . ."

I paused. Took a breath deep enough for both of us. Continued.

"I don't quite know how to break it to you, but, as a writer, you've sort of fallen off the charts."

This disclosure, too, elicited no response—which is not surprising. Rejection, for any writer, is a tough pill to swallow.

"It's good stuff," I quickly added. "Don't get me wrong. It's just that some of your material is pretty dated, now. And Amazon.com's never heard of you. And if you don't mind my saying so, it seems that, despite your fine work, young people are more estranged from the natural world than ever."

Took the news really well, he did. Near as I could tell, with detachment.

"So listen, Dallas—do you mind if I call you Dallas?— I've got a proposition to make. You see, I've got this book under contract. And it's going pretty well . . . or, at least, I thought it was going pretty well.

"Then I read a copy of your book *Summer*. And I read the chapters entitled 'The Summer Afield' and 'Things to See This Summer' and 'Things to Do This Summer.' And I realized that I have only four days left on the calendar for this project. And that I really left a lot of ground uncovered—suggestions and avenues of exploration that you've so admirably outlined in your book."

I knew I was laying it on pretty thick here. But maybe the dead are not above flattery.

"So I was wondering whether, maybe, you and I could collaborate a little. I could excerpt some of your stuff and rework it to appeal to a more contemporary audience. You'd get your name back on the charts. We'd kindle a bit of new-found interest in people who have never heard a marsh wren's song or watched an osprey tower.

"We call fish hawks 'ospreys' now," I added.

The naturalist didn't respond. But a mockingbird, which had until this moment been sitting quietly atop a nearby cedar, abruptly flew to the ground and started strutting nearby, executing a smart open-wing display that carried it right across the naturalist's marker.

Mockingbirds are a common resident species throughout New Jersey and, now, north into New England. But in Sharp's time they were decidedly less common. Tenacious singers, mockingbirds incorporate the songs of other birds into repetitive ensembles that they use to sing rings around their territories. They are, in essence and practice, experts in plagiarism and the field of intellectual property rights.

"You didn't tell me you had an agent," I said to the plaque.

The bird reversed course and walked over Sharp's grave again before turning, facing me, and lifting and opening its wings in a hard to mistake gesture.

"No, the royalties aren't that grand. It's a low-budget effort."

Undaunted and unperturbed, the bird lifted off and flew to the top of the tall monument that dominated the Sharp family plot.

"I suppose that means he wants to be primary author."

The dapper gray mimid made a barely audible two-note murmur and returned to the cedar tree, where it began to preen.

"Okay," I said. "Agreed. But only on this chapter, and only if we agree to use the new bird names as established by the seventh edition of the AOU checklist."

The mockingbird never even blinked. It was, after all, a small point.

"Let's get to it," I said, reaching for my daypack, pulling out a copy of *Summer*, and flipping to my first Post-it note. Hoping that Irene Ferguson, president of the Mauricetown Historical Society and the one who had loaned me the book, wasn't going to stop by to pay respects to her late husband and see how I was mistreating an object from their archives.

"On page two," I said to the plaque, "you talk about outfitting young observers for a 'tramp' outdoors. I like it. It's got the nostalgic feel, but that's what we're trying to capture here. A return to good, healthy, virtuous values. This will play well with Evangelicals, who make up a big book-buying block."

Having a trained minister, as Dallas Sharp was, as a coauthor wasn't going to hurt sales either.

"Now, here's the problem. After your suggestions regarding 'stout, well fitting shoes' and 'clothes that will not catch the briars' you urge young readers to 'swing out on [their] legs and . . . do ten miles up a mountain-side or through brush.' And you say, 'If at the end you feel like *eating up* ten miles more, then you may know that you can walk, can *tramp*, and you are in good shape for the summer.'"

The noted author seemed not to see the problem with this.

"Uh, see, I don't think we're going to find many kids these days who are going to walk five miles (much less ten or twenty) without an inhaler. And most parents don't let their kids do anything outdoors alone, without adult supervision and a wad of consent forms and contact information and a data bank's worth of emergency medical information exchanging hands.

"Any parent who lets his kid just bushwhack, alone, into Bear Swamp would be handcuffed, and the kid would be in the back seat of a sedan licensed to the Division of Youth and Family Services.

"What do you say we trim the mileage back to, say, two and pitch this chapter to organized youth groups and focus on trail etiquette and developing good group problem-solving and GPS navigational skills?"

The mockingbird muttered something inaudible, raised its tail, ejected a short, to the point, whitish squirt.

"I know. It's going to mean kids are going to see and engage less. Wildlife learns to avoid established hiking trails during peak people times. And discovering and engaging nature on your own is much more gratifying than doing it by committee. But everything is organized these days, and adults who could serve as mentors are too committed to go one-on-one with kids.

"Think about it," I concluded, realizing that there was a lot of ground to cover and we had to move on.

"Okay, on to equipment, the stuff in 'your tramping kit' as you put it. The pocketknife is a big nonstarter. I know that, in your time, pocketknives were only slightly less common in kids' pockets than lint (heck, I got my own first Swiss Army knife when I was nine). But we're not far from the point where knife ownership is going to have a minimum age requirement and people are going to have to pass background checks in order to purchase and carry.

"As for 'string,' well, they won't even allow passengers to carry string on an airplane anymore. So long as a kid is carrying enough granola bars and trail mix, no one is going to have to set a snare to catch a rabbit for lunch or jury-rig a fishing pole. Hate to tell you, but we either ditch the string idea or get used to having our names on the watch list of the Department of Homeland Security.

"But 'camera'! Right. Every kid's got a cell phone that captures images these days. It will complement your challenge for kids to 'see and watch sixty species of birds' over the course of a summer. They can see them, snap pictures, download the images, and get somebody else to identify them.

"Heck, in a couple of years there will probably be computer programs that will identify birds for them. Field identification will be about as archaic as changing a typewriter ribbon or shooting marbles."

The mockingbird turned his back on me. Not that I blame him. Shooting marbles was a lot of fun.

"Moving on, in chapter 11, you recommend that young students of nature 'select some bird or beast or insect that lives with you in your backyard or house or near your neighborhood, and keep track of his doings all summer long, jotting down in a diary your observations.' You identify this as your 'training exercise for patience and independence.'

"The backyard context is great. It's about as far as kids get these days. They can get their images. Post them on YouTube. Do a visual essay and probably fit the whole thing in between swimming and fencing lessons and sessions with assorted online games. It will take them hardly any time at all. And Mom won't have to drive.

"Fact is, however, that, as an exercise in patience and independence, this might not be as important as it once was. Kids today have developed almost limitless patience, and their capacity for independent activity seems staggering. I know kids who spend every free moment sitting in a chair playing World of Warcraft online."

The mockingbird looked at me and blinked.

"It's a lot to explain," I said. "Let's just say that, when adults stopped letting kids go outside, kids found a parallel universe to engage them, one that pushes the buttons on some human senses but cuts the corners on others. Speaking of which . . .

"About your suggestion that kids 'browse and nibble in the woods . . .'

"Look," I said, first to the plaque, then to the bird, "I know that smell and taste and touch are bonding and that knowing what to eat and what not to eat is the ultimate intimacy with the natural world. Heck, I used to eat wild cherries until I was sick, dug up and chewed enough sassafras root to wear the enamel off my teeth, and cannot imagine a childhood deprived of the sweet taste of white oak acorns and the bitter taste of red and the wisdom to know which was which.

"Discovery is what childhood is made for, and memory is what adulthood is made of. But we're dealing with a generation of kids who grew up with the telephone number for the poison control center programmed into the phone and parents who didn't even trust the safety of food they bought in the store. Encouraging kids to graze off the land is going to get us sued."

The mockingbird leaned over. Plucked a small, grayish juniper berry off the bush. Bolted it down. He plucked another. Knocked that one back, too.

"I thought you were a temperance man," I said to the sober Dr. Sharp, who once sought to run for the United States Senate on the Democratic ticket, employing in his idealistic platform a prohibitionist plank. He was soundly defeated. My grandfather Dunne would have cheered.

"One last thing," I said. "As for encouraging budding young naturalists to consider 'spending some time this summer on a real farm,' well . . ."

I looked around the cemetery. Three hundred and sixty degrees around. There were houses. There was a woodland. There were two baseball fields (empty, of course). There was, in what was once known as a "farming community," a notable dearth of farmland.

"I really don't know how to break this to you, but farms are getting about as hard to find as temperance wagons. We're still pretty lucky here in Cumberland County. We still have working farms. But more and more, small working farms, operated by families, are becoming so rare that they are closer to falling under the jurisdiction of the Department of the Interior than the Department of Agriculture. I doubt that one kid in a hundred has had greater 'contact with growing things in the ground' than the grass in the front yard that they aren't allowed to play on because it's been treated.

"They might, in their lives, as part of a school field trip, experience 'the mere visit' to a farm that you allow is not enough. But hoeing, gathering, feeding chickens, driving cows to pasture, baling hay, and 'all the other interesting experiences that make up the simple, elemental, and wonderfully varied day of farm life' well, unless Nintendo develops a video game called 'Down on the Farm,' I doubt that many kids are ever going to get a taste of this."

Too bad, really. As mentioned, I spent a couple of summers on a farm. I will go to my grave remembering them as the two best summers of my life.

The mockingbird had stopped indulging, in fact, had flown off, taking a perch on the peak of the church roof, from where he could no doubt see, with surrogate eyes, the forests, the marshes, the river. In his book *The Fall of the Year*, Sharp reminisced about the region's ecological trinity. "More and more, as the years lengthen," the naturalist reflected, "do we find ourselves longing—for the pine barrens, for the past reach of the marshes; and were my feet free this summer day, they would run with my heart to the river— not the mountains; to the river, the Maurice River, where the bubbling wrens build in the smother of reed...and where this very day the pink-white marshmallows make, at high noon, a gorgeous sunset over miles of meadows. I love and understand those great, green levels of marshland as I shall love and understand no other face of nature it may be."

And now, on the far side of summer and near the end of this book, I find in this description, written one hundred years ago, more engaging wisdom than I have housed in the whole of mine. Insight, too.

What an old man reminisces about, in the autumn of his

years, is not what he has engaged and understood. It is what he has engaged and loved.

The wrens, just as Dallas described them, are still there. The marshmallow, too. And the golden-eyed fish hawks that Dallas Lore Sharp loved so much still mantle the sky on crooked wings and fill the air over Mauricetown with their piping whistle.

Very probably the hovering osprey that greeted Linda and me on the first day of this project was a descendant of birds that a young Dallas Sharp marveled at as he "swung out" to engage the world around him. Were Dallas Lore Sharp to "tramp" the earth again, he would find the woods and marshes and river and the region's wildlife just as engaging and nurturing as he remembered them. It's the human culture he would find alien.

The first heavy drops of rain, spun from the moisture drawn up from tropical seas, were beginning to fall. Enough to make me fear for the book in my hand. Enough to make me want to conclude our business meeting.

"Pleased to make your acquaintance, Dr. Sharp," I said. "We'll see if we can get a little more play out of your book; see if maybe between your words and the region's charm we can't find a few young tramps, or tramps at heart, who'll want to come here and discover what you and I discovered.

"But I'm going to leave the tasting part out. Never get it past my editor or Houghton Mifflin Harcourt's legal department."

I waved to the mockingbird (who ignored me). Plucked a juniper berry off a lower branch, bit into it, and inhaled—

through my open mouth—a breath of gin-tinged air. Thinking that it might be the worst possible luck to be a kid born today. Realizing, suddenly, that the present estrangement of kids and nature probably wouldn't have surprised Dallas Lore Sharp at all. Troubled, yes. Shocked by the level of estrangement, undoubtedly.

Surprised, no. After all, he wrote his book in 1911 and clearly saw the beginnings of estrangement then. Things haven't changed so much as they have gotten worse, with more and more people not only uncomprehending of the natural world but alienated from it by the pattern of their lives.

As I made my way to my car, past the graves of sea captains, cedar miners, farmers, 'erstermen, baymen, people whose lives were wedded to their environment, it occurred to me that, for many people living today, their first real contact with the earth may very well also be their last. Then, looking over at the gentleman, now shirtless, who was finishing preparations for the impending graveside ceremony, I watched as he lowered the concrete burial chamber into the ground, realizing, sadly, that, in this day and age, people are being deprived of this final contact with the earth, too.

Being "returned to the earth" isn't what it used to be either.

The rain was falling harder now. It would be a good afternoon for writing. Lousy day for a funeral, though.

CHAPTER 12
The Last Kumor of Thompsons Beach

Labor Day Monday, and the Wawa was hopping once more, jammed with people trying to squeeze one last outing out of the summer—and for the first time kids were out in numbers! It was as if a whole generation had awakened to the realization that summer had all but slipped away.

Other than the surge of school-shocked youngsters, it was a typical holiday crowd. Lots of shore tourists on the rebound. Lots of roofers, painters, plumbers, and laborers once more transformed into fishermen, backyard chefs, and horseshoe-pitching champions by the magic inherent in a holiday weekend.

Growing up, I was wryly amused by Labor Day—of all the legislated holidays the most curiously contrived. A day set aside to honor those "who from rude nature have delved and carved all the grandeur we behold," as Peter J. McGuire, cofounder of the American Federation of Labor and possible Labor Day founder, explained it. But the holiday was celebrated not by flexing muscle, as the name implies, but by doing just the opposite.

When I came to understand that Labor Day had less to do with actual work and more to do with organized labor, the holiday lost some of its oxymoronic suggestion. The first Labor Day was, in point of fact, *organized* by the Central Labor Union of New York City on September 5, 1882, and celebrated, as it is today, with speeches and a parade. The one-day job action was intended to draw public attention to the plight of working men and women and to demonstrate the political clout of organized labor.

But the day, and its message, didn't get real traction until the Haymarket Square Riot on May 4, 1886, when President Grover Cleveland dispatched troops to Chicago to put down a strike by railroad workers. The bloody confrontation brought workers' rights to the forefront of public attention, resulting in pro-labor legislation and a congressional decree that the first Monday of every September would be recognized as Labor Day, an official federal holiday.

More than one hundred years later, Labor Day is still celebrated with parades and speeches by politicians who praise the contribution of working men and women (particularly during election years). Most Americans just use it as an occasion to take off for one last day of picnics, barbecues, and getting out on the water.

I was going out on the water, too. With Captain George Kumor of Heislerville. But, unlike so many working Americans, we'd be celebrating the holiday by actually working.

Working on the water. Working toward some political and philosophical common ground between those whose lives engage the natural world and those who strive to understand and protect it.

The foundation of our alliance was a mutual love and respect for the natural resources of the bayshore. Cementing it was a mutual enemy: the societal estrangement and compounding challenges that threaten the region's historic, cultural, and biological integrity.

It was a lot to fit into a sixteen-foot boat. And resolving all the problems facing the region was not going to be as easy as simply declaring a holiday.

This time I got the twelve-ounce coffee. I didn't know how long we'd be out and figured it was wise to monitor my fluid intake. Reaching my car, maneuvering around boat-towing SUVs and the first wave of homebound traffic, I turned right out of the Wawa, heading south once again on old Route 47. Past the turnoff to Dorchester. Past the turnoff to Leesburg. Left onto Route 740 just past the official green highway sign directing travelers to Heislerville, Matts Landing, and Thompsons Beach.

If the names sound familiar, it's because you've navigated this route before; Thompsons Beach was the place Linda and I went to photograph shorebirds in the opening pages of this book.

And if the name Kumor also strikes a chord, it's because

you've been introduced to him before as well. Captain George was the waterman whose boat hove into view on that beautiful Saturday a summer's span ago.

Putting in each day at Kumor's Ditch—a man-made waterway slicing through the marsh that was once the Kumor family's holdings—he heads out to the bay to drop his net and pull his traps an evening shadow's span from the rubble of the town he grew up in.

George Kumor. A fifty-five-year-old native bayman, in a small boat, plying an ancient trade in a mercurial world that increasingly seems run by experts instead of the wise.

OF TWO GIVEN ALTERNATIVES, ALWAYS CHOOSE THE THIRD

We'd agreed to meet at 8:00 to catch the tide. I was early, George late (by the clock, not the tide). During the interval, I had time to mount the observation tower that had been erected as part of the wetlands restoration project, which changed the face of the bayshore, resulting in, among other things, the relocation of Thompsons Beach residents and removal of the beach's structures—including the piers and docks once owned by the Kumor family.

When I'd first navigated this stretch of road, back in the late 1970s, it was bracketed by diked salt hay meadows. Along the road, you'd find breeding saltmarsh sparrows and nesting northern harriers, birds that thrive in high marsh habitat.

At the end of the causeway was Thompsons Beach, or what remained of it. Even in 1979, George Kumor's boyhood town was leading a tenuous existence, its beach pushed back by the encroaching waters of the bay. By the mid-nineties, when one of New Jersey's major public utility companies came in with offers to buy up the town and surrounding land, most people were at least willing to listen.

The situation and the proposal went something like this. In the 1970s, Public Service Electric and Gas, one of New Jersey's most respected and responsible corporate citizens, built a nuclear power station at the headwaters of Delaware Bay. It provided power to the region. It created jobs and reduced taxes for the people living in the municipality of Alloways Creek. It seemed like a good thing.

But over time, it was noted that Delaware Bay's celebrated weakfish population was failing, and the nuke plant was implicated, the suspicion being that large numbers of young weakfish, and other marine life, were being sucked in and killed by the plant's water-cooling system. After a protracted period of legal and political wrangling, PSE&G was faced with closing their facility or replacing it with one that was more fish-friendly—choices that would cost the power company hundreds of millions of dollars in lost revenue or construction fees.

Then some clever, and motivated, people came up with a third alternative. Instead of eliminating the damage caused by the reactor, why not mitigate for it by bolstering the bay's productivity? It was well known that many marine organisms spend part of their life cycle in the creeks and tidal estuaries of Delaware Bay. If PSE&G were to construct inlets and channels restoring daily tidal flow to marshlands diked for the production of salt hay, it would, presumably, increase overall salt marsh productivity and offset the damage the reactors had done to weakfish and other marine organisms.

What's more, the project would reduce mosquito populations by increasing tidal flushing in upland breeding areas, purchase thousands of acres of buffering upland to protect the bay from future encroachment, and incorporate a network of trails and viewing platforms so that visitors could enjoy greater recreational access.

Platforms like the one I was standing on. Inlets and channels like Kumor's Ditch.

The best part was that the mitigation proposal would cost only a fraction of the price tag for a new, fish-friendly reactor.

Win, win, win, win, WIN!

And so the very ambitious Wetlands Restoration Plan was designed. Offers for the purchase of land from farmers and residents were tendered, and deals were made. Many landowners in this economic backwater region were pleased with the money and the settlement. A few, like the Garrisons, said "thanks, but no thanks," and a few hold-outs, in places like Thompsons Beach, ultimately surrendered to the power of the rising tide and eminent domain. The project took three years and was completed in 1997.

With what result? More than eight thousand acres of diked upland marsh were returned to tidal flow. An additional six thousand acres of buffering upland was acquired to protect the wetlands from potentially damaging encroachment. The high-tide line was pushed inland, and forested necks of land inundated, with the result that many acres of bordering forest were killed by saltwater intrusion.

Breeding populations of northern harriers, black rails, saltmarsh sparrows, and eastern meadowlarks declined. They were replaced by increased numbers of breeding ospreys, clapper rails, and seaside sparrows.

And the weakfish population? It crashed.

Go figure.

George's vintage pickup hove into view at 8:10. He was already wearing his trademark headgear—a makeshift French legionnaire's cap consisting of a neck-protecting bandanna held in place with a sun-bleached baseball cap. While George backed his boat into the water, I got into the bayman's uniform he'd handed me: bibbed yellow coveralls and matching oilcloth jacket. It was clear I wasn't just a passenger on this Labor Day outing.

THE E. F. HUTTON OF THE BAYSHORE

George Kumor lives with his wife, Anna, in Heislerville in what they both describe as "a fisherman's house." What this means is that you don't have to take your boots off when you walk inside (but people still appreciate it when you do). His commute to work is about three minutes, largely because, towing a boat, he drives slowly.

He bought his first boat in 1967, at the age of fourteen. He put in at his father's Thompsons Beach dock. Dropped some crab pots just off the town. Sold his catch locally. Give or take a few years at private schools and a smattering of college, he's been doing it ever since.

Plying the inshore waters of Delaware Bay right off the town he grew up in. Selling his catch to the people of the village. Spending his free time going to meetings, serving on councils and committees whose focus or jurisdiction includes the bayshore environment George calls home.

Accounted one of the "village elders," he allowed—or let slip—that he is known locally as the E. F. Hutton of the Bayshore. If the meaning of this escapes you, then you somehow avoided the Wall Street brokerage firm's advertising campaign whose tag line was "When E. F. Hutton talks, people listen."

I can't swear to the universal attentiveness of listeners. I can tell you that just about everyone who has had anything to do with the Delaware Bayshore in the last twenty years knows George Kumor.

We'd met a decade earlier—or at least this is when I remember meeting him—when, after the "restoration," he started putting in at the newly cut Kumor's Ditch. He was at the boat launch one spring day when I came down looking for shorebirds. He was just pulling his boat from the water. The ditch was almost clogged with horseshoe crabs—many dead and dying.

One of the unforeseen consequences of the restoration project was the redirection of breeding crabs off the beaches and into the new man-made cul-de-sacs. Many thousands of animals were stranded in the man-made killing fields behind Thompsons Beach. It didn't help crab populations. It didn't help the dwindling shorebird numbers. It muddied the water when blame and responsibility were being apportioned over the decline of horseshoe crabs. Horseshoe crab harvesters were quick, and to some degree correct, to point the finger at the restoration project and say that it, not they, was responsible for the decline in crab numbers.

The truth? Every little bit hurt. And the first casualty, in any conflict, is usually the truth.

So it was during this period of conflict and cross-accusations that I met the curious man with the makeshift legionnaire's hat. He was about six feet tall, of solid but average proportions, but he moved with a controlled quickness that suggested physical strength, or anger, or both.

His eyes, set behind square-framed glasses, were pale and appraising. His manner not unfriendly but nevertheless direct. If you passed him on a subway platform, in a business suit, with a newspaper folded under his arm, it would seem as natural as nine to five. In his bibbed overalls and T-shirt (and except for the legionnaire's headgear), he seemed the picture of a commercial fisherman, someone whose interests in the bay were, I assumed, contrary to mine, and whose appreciation of the bay was, most certainly, tainted by economic self-interest.

I was wrong on both counts. First, he was not a commercial fisherman. He was a Delaware waterman, a bayman. "Commercial fisherman" was a label "environmentalists" put on him.

Second, his knowledge of the bay and the workings of its

environment was broad and deep, not economically shallow, and he was not unequivocally on the side of horseshoe crab harvesters. He admitted, frankly, and much to my surprise, that harvesting crabs was not the deep-seated cultural tradition that the harvesters presented it as; that there were only a handful of people harvesting crabs before prices jumped to a dollar a crab.

But George was equally critical of the birders and environmentalists, identified to me as "you people." Researchers who, like the blind men and the elephant, came down to the bay to study a narrow slice of its ecology and make sweeping decisions and pronouncements about how things should or should not be done.

" 'Minutemen' is what we call them," George said with an expression that fell just short of a smile. "People who visit the bay. Spend a minute looking around. Leave with all the answers."

In this regard, he manifested the classic insular smugness that is endemic to local people everywhere. The "we live here and we know better than outsiders how things work" attitude. The head-in-the-sand defensiveness that causes local people to get run over by outside interests time and time again.

My response, at the time, was to adopt the cool air of detached superiority that academic training confers upon a person and say nothing.

As strategies go, it was about as flawed as putting my head in the sand. Because the next subject George brought to the floor was the environmental community's lack of involvement. How we talked a good war, but when it came down to going to meetings to, say, fight sewage discharge from the prison, he never saw any of "you people" (meaning me) there.

It was my time to be defensive. What George was saying was true. And explaining that my organization's mission focus and my field of expertise weren't water quality matters, while true, seemed a weak defense.

The guy had gotten under my skin all right. What impressed me greatly was not only his knowledge and acumen but his obvious passion and concern.

We environmentalists like to think that we have a patent on concern. This presumption, coupled with our narrowly defined focus and academic superiority, makes us insular and smug.

Anyway, that's how George and I met. In a funny sort of way, I think we both respected each other and maybe even trusted each other. The challenge was whether we could understand, much less accommodate, each other.

Gear stowed and small talk concluded, the bayman motioned for me to take a seat in the bow; he, tiller in hand, took the stern. Between us lay an assortment of coolers, buckets, peach baskets, 360 feet of drift net, and a lot of other accouterments that have proven their usefulness over the course of George's forty-plus years on the water. Everything not recently moved into the boat was covered by a gray patina of Delaware Bay blue mud. Under George's boots, a puddle was already growing—demonstrating the usefulness of one of those utilitarian accouterments, a bailing can.

As we eased into the channel, we crowded an adult black-crowned night-heron standing on the bank, who nevertheless allowed us to get unaccountably close. His name, George said, was Harold.

"I told him you were all right," George informed me. "I gave him a hand signal."

It would be very presumptuous on your part to say it wasn't so.

SHORT RUN TO THE BAY

We stopped first at a boot-slickened patch of mud where George distributed yesterday's leftover bunkers for the gulls. This effort is part of a pilot project, initiated by George, to keep the larger gulls fed so that they don't compete with northbound shorebirds for horseshoe crab eggs. He has been trying to interest the people "from the state" and "environmentalists" in supporting this effort, but with no success.

The reason, of course, is skepticism. Skepticism that such a simple, homegrown effort could have a measurable impact on the feeding success of shorebirds.

I might have pointed out to George (but didn't) that horseshoe crab eggs are considered too small for larger gulls to feed on, especially since those eggs are no longer found in superabundance. The smaller laughing gulls, yes, they did and do compete for eggs, bolt them down like pinhead-size gumdrops. But this high up in the bay, this far from breeding colonies, most of the gulls are nonbreeding subadult herring and great black-backed gulls—birds that forage on dead and dying horseshoe crabs, not eggs; food that offers a large gull more energetic bang for the foraging buck.

Yes, I'm skeptical, too. But living on the bayshore as I do, my skepticism is tempered by an undermining truth. I don't know for sure. My background in this matter goes back twenty-five years, only three of which were dedicated to the study of shorebirds. George's "pilot study" goes back the length of his working life in a field of study whose cumulative wisdom was amassed by generations of baymen. The study of men and how they affect and are affected by the bay.

At the very least, and unlike some projects that "experts" have conducted over the years, feeding unused fish to gulls probably wasn't going to hurt the study population.

And it is very much in keeping with the ethic George grew up with, which is "waste nothing." The fish George had taken from the bay, he was returning to the bay. It is the ethic of people who strive to live in harmony with their environment.

"You weren't at the meeting last Thursday," he said, suddenly.

He was right, again. Maurice River Township holds its council meetings every third Thursday, and one of the items on the agenda had to do with ecotourism. George had invited me to serve on the committee.

"I know," I said. "I had to go to the purple martin roost that night. It looked like the birds were building up for their big exodus, and I didn't want to miss it."

George didn't respond to the excuse, and my guess is

that's because he recognized it for what it was, an excuse. Fact is, I have an abiding (and somewhat famous) dislike of meetings, but as a senior member of New Jersey Audubon's staff, I am obligated to attend a lot of them. It's my daytime job. In the evening, in the name of sanity, Linda and I pull up the drawbridge and let the world go to hell by itself for a time.

It is George, a fisherman, who engages the natural world all day and then at night goes to meetings where the decisions concerning his bay and his environment are made. It begged the question who was the real "environmentalist" in the boat?

We reached the mouth of the bay with startling suddenness. One moment we were floating in a tidal creek, the next we were on open water. An osprey, perched on a nearby platform, whistled its annoyance and fled.

I looked at George and my eyes asked.

He looked back and his eyes said.

"Yep. Bird didn't know you."

Hell of a thing. I spend my whole adult life working for the welfare of birds, and my diplomatic credentials aren't even recognized. George, on the other hand (and as another fisherman), gets treated as a colleague.

"Does the osprey have a name?" I asked.

"Doudah," he said, nodding.

It was a beautiful day. Clear, calm, warm. Surprisingly, despite the holiday, there were few boats to be seen. One other commercial boat, several distant sport fishermen.

"What kind of summer has it been catch-wise?"

"Terrible. The worst ever."

Slowing the boat, George abruptly stood, faced the shore, made the sign of the cross, and bowed his head. He has family buried on the family holding. He conducts this ritual every time he goes out.

After finishing his prayer, George finished his fishing report. "It's so bad that this year they had to cancel the annual weakfish tournament. First time ever."

With this harsh disclosure hanging in the air, George started letting out his gill net, allowing the wind and the tide, as much as the outboard, to do the work for him. We'd haul net on the way in. See what modest fortune could be seined from the sagging fortunes of Delaware Bay.

Dead Sea and Designs Gone Awry

We reached the first of George's crab pot markers. He idled back on the engine, grabbed the float with a gloved hand, and brought the trap in. It was, in essence, the same procedure I'd observed on Tom Pew's boat except for three things. It was done slowly, as by a man engaged in a craft, not hurriedly as by a worker on a production line. It was done by a single man, not a team. It was done completely by hand. On Tom's boat, the pots were grabbed with a hook and drawn from the water by a machine. On George's boat, the hooking, the hauling, the emptying, and the rebaiting were all done by a single bayman. A single man. And it was all done by hand.

Fishing is an ancient trade, one long predating the industrial age that spawned the labor movement. It was an age when labor was done by tradesmen, not workers. What they engaged with their hands belonged to them, not to the owner of the factory, who then sells what the workers produced to middlemen, who resell it to distributors, who resell it to consumers.

George sells his catch directly to the people of the village. When he has a catch.

We were operating in two fathoms, twelve feet of water. In six quick pulls, George had the trap out of the water and balanced on the side of the boat. The sound of water running rivulets from the cage gave way to the skittering sound of crabs coming to terms with a sudden change in their fortune and their environment.

"This is bad," George said quickly.

"What's bad?" I said, trying to see with my untrained eyes what George could see with his waterman's eyes.

"The dead crab," he said, giving the trap a shake, lending animation to half a dozen blue crabs and distinction to one.

"Why?" I asked.

"Lack of oxygen," he said.

"Why?" I asked again.

"The wetlands restoration project," he replied. "It's used up all the oxygen."

"How?" I wanted to know, and George, who wanted me to hear his take, told me.

The restoration project, as you will recall, was intended to increase marine productivity by adding more land to the tidal cycle. More wetlands equals more wetland productivity. More wetland productivity means more recruitment for creatures like weakfish, which spend the younger portion of their lives in the estuary and then, as adults, live in the bay.

As George explained, everything went pretty much according to plan at first. The old salt hay farms and uplands were flooded. High marsh quickly reverted to tidal marsh, and the *Spartina alterniflora,* the cord grass that flourishes in flooded marsh, grew thick and lush. If the measure of the success of the project had been green grass, not increased weakfish populations, the program would have been a whopping success.

Then, according to George, things began to break down, literally. All the organic matter from the plants killed by the inundating tides began to decompose and get flushed into the bay. This nutrient-rich soup made the algae bloom, which depleted the oxygen in the water, driving small fry that used to hide out in the shallows into deeper water, where they were clobbered by larger, predatory fish, most notably striped bass, whose population exploded.

It was plausible analysis. The decline of weakfish in Delaware Bay did continue after the mitigation project was completed. What's more, according to state fishery biologists, the decline does not seem related to any dropoff in weakfish productivity but, as George suggested, to survivability.

But correlation is not the same as cause and effect, and factors affecting fish populations are many and interrelated. Weakfish populations along the entire Atlantic seaboard, not just Delaware Bay, are declining. This decline correlates with rising seawater temperatures, although how this affects weakfish populations is uncertain.

Maybe higher water temperatures advantage fish that prey on weakfish. Maybe they negatively impact populations of fish that weakfish feed upon. Maybe they, as much or more than nutrient enrichment, accelerate the growth of algae, which, as George notes, results in the deoxygenation of surrounding water.

Only two things are certain. The decline of weakfish in Delaware Bay has been more rapid and dramatic than elsewhere. And there was a dead crab in George's trap.

Quickly, deftly, George shook the live crabs from the trap into a waiting barrel and rebaited the pot, tearing the tails off several bunkers he would use for bait, tossing about one third of each fish overboard.

It was then that I noticed the gulls. Several score herring gulls that had gathered all around George's boat. George's minions. Feeding the people of the village is not the only charge that George has assumed. He has also taken upon himself the feeding (if not the care) of a number of subadult and adult gulls (whom he knows and calls by name). Among their ranks, now late in the season, were a dozen or so dark brown juveniles.

"Chocolate drops," George called them. These young birds were just picking up the scavenging trade.

And I know that there are some people, many of them environmentalists, who are going to read this line and believe right down to the core of their educated and environmentally aware beings that what George is doing is a disservice to wildlife, constituting an unwarranted intrusion into the lives of wild creatures who should be honing their foraging skills.

Look. These are professional gulls, and gulls are, by trade, opportunists and scavengers. If one kind, creature-friendly bayman's by-catch constitutes a food source for half a hundred gulls, it is hardly an ecotipping matter.

Not even close to converting thousands of acres of high marsh to tidal flow.

And from factory fishing ships operating in international waters to open landfills, to untreated sewage, to French fry–brandishing tourists, gulls have been taking advantage of human by-catch, by-waste, and unmindfulness for a long, long time.

As I watched, one adult herring gull landed on the stern of George's boat. George reached into the bait bucket, broke a bunker in three pieces, and offered the bird not a tail but a nice, flesh-rich, trap-worthy cut of menhaden.

"That's Matilda," George explained. "She won't eat the tails."

HAPPY LABOR DAY

In slow succession, George pulled the other traps in the first string. All held crabs, although none held more than a few, and, fortunately, no more of the crabs were dead. We motored over to George's next string, but instead of lining up immediately on the first marker, George invited me to open a plastic bag he'd set in the bow. In it were two sets of gloves: cotton undergloves and a rubber outer.

The exercise had just stopped being academic. I was about to step out of the ranks of the bourgeoisie and join the proletariat.

George grabbed the first float and handed the line to me. I pulled, off balance, realizing quickly that a change of tactic or a trip to the chiropractor was in my future.

Repositioning my feet and facing forward, I hauled line, hand over hand, feeling the good strain of weight on muscles, feeling more exhilarated than I've felt since . . .

Since bailing salt hay in July.

The trap broke water, and, deprived of its buoyancy, the weight of the thing tripled. Unprepared, I struggled to get the trap onboard and was grateful when George lent a helping hand and finished the process.

"Let me do the next one by myself," I heard myself say.

"Okay," he said, and I did.

Grabbed the float with a one-hand grab. Hauled until the trap broke water. Put my back into it, bringing the weighted—and crab-bearing—cage aboard. Puzzled over and solved the mystery of the bungee cord that held the trapdoor closed. Turned and shook the trap, watching as half a dozen blue crabs surrendered to gravity and fell into the waiting bucket. Inverted the trap (realizing that I'd failed to dislodge the old bait, which would now have to be pulled by hand).

Paid for my mistake. Snapped the tails off two semi-frozen menhaden, stuffing them headfirst into the trap's bait bin and tossing their tails to two gulls chosen randomly (no doubt screwing up the pecking order). Capped the bait with a fish jammed in sideways (like I'd seen George do).

"Turn it over quickly," George counseled. "The bait'll stay in."

I did. And it did.

I pulled four more traps. Enough for the pleasant ache to become a real ache and for me to appreciate, somewhat, the amount of labor that goes into being a Delaware bayman with traps to pull and a village to feed.

George's point exactly.

"I've got to take some notes," I said, I lied, stripping off the gloves. Grabbing my notepad. Stepping back from the ranks of workers and becoming, once again, one of the planet's superior creatures—the ones whose lives are indentured to ideas affixed to paper, as opposed to those whose hands engage real, living things.

But in my soul, I felt a new smugness vying with the old. For a few minutes, I'd been in a bayman's boots and had a taste of the bayshore's tradition. Not near enough to instill a bayman's wisdom. But enough to instill a sense of kinship and peer beneath the veil of smugness that conceals a bayman's pride.

Murky Waters

We finished the second set of traps and went on to the third—all of them set in the shallow waters two to three hundred yards off Thompsons Beach. Even to my untrained eye, it was obvious the water was darker, murkier here. The trap George pulled was encased with gluelike algae, and it held no crabs. Neither did the next or the one after.

The traps we pulled with crabs were near the mouth of the ditch, I thought, I knew.

"The water here is deoxygenated. Dead."

"Take a look around," George suggested. "See any boats?"

I did and I didn't. Except for the same distant sport-fishing boats out on the horizon (manned by people who were there because of a date on the calendar, not the presence of fish), the bay was empty. In fact, except for the single commercial crabber seen earlier, I realized we'd encountered no one.

"Notice how he [meaning the other crabber] quit and went in?" George asked. "It was costing him more to be out here than he was making, so he went back to the dock."

George, with a smaller boat and no mate to pay, stayed out. Besides, he had the people of the village to feed. And he had a writer in the boat. And he had, as I did, an agenda.

When I'd first approached George about letting me go out with him, his response was diffident. He wasn't hostile. But he was more suspicious than amenable—not that I blamed him. In the history of the bayshore, when people from outside come asking for something, it is usually a thing that comes at your expense.

And no, living in Mauricetown does not automatically make me a part of the bayshore community, nor does it necessarily endear me to people in places like Heislerville. Mauricetown was where ship captains lived, the bourgeoisie. Heislerville was where the baymen, and shipyard workers, lived. The proletariat.

George had Linda and me pegged as people who lived in

one of the impressive Victorian homes on the riverfront. It helped my case when he learned that we lived on Second Street and that our small home was built by a ship's blacksmith.

But more and more, year by year, people like Linda and me are becoming the future of the bayshore, and people like George, his wife, Anna, and Bill Garrison and Tom Pew are becoming its past. After decades of isolation, this overlooked corner of New Jersey has finally been "discovered." Its resources are being coveted and contested. Its four-hundred-year-old tradition assailed on social, economic, and political fronts.

One of my reasons for writing this book was to try to portray and preserve something of the unique and dwindling heritage of this little-known region. When George was assured of this, and that I wasn't simply looking for a "folklore figure," he agreed to take me out. It turns out his interest and mine were one. Captain George Kumor, one of a dwindling number of men still working the waters of Delaware Bay, sees more clearly than I do the end of four hundred years of tradition and worse. The ecological erosion of one of North America's last great natural areas.

That's what unites us. What divides us is our relationship to the bay.

George works the bay. I study the bay.

George's strength comes from wisdom. I come from a discipline that puts its stock in "experts."

George, a craftsman, engages his work and his world from start to finish, beginning to end. I'm a specialist, among other specialists, in an organization that survives and distinguishes itself from other organizations by its special focus and expertise. We do our special thing. We trust other state, federal, and private organizations do their special things, be they marine ecology, habitat management, land acquisition, or water quality. At the end of this environmental assembly line, we have a nice, healthy, flourishing environment for everyone to enjoy.

At least that's the plan.

And here we were, two earnest men in our mid-fifties, sitting in a small boat, in a big bay, divided, not by a common interest but by our relationship to the environment.

Hands on versus hands off, and the biggest problem is that ours were not the only hands in the game. There are other people with designs upon the region. People with ideas. People with ambitions that will in the years ahead have direct and indirect impacts.

Development pressure along major watercourses whose waters sweeten and nourish the bay.

Plans that call for the waters of Delaware Bay to be used to dilute waste and chemical discharge.

The development of aquaculture facilities to augment or supplant the commercial harvest of native fish species.

The erection of offshore wind turbines to fuel the region's growing energy needs.

The ongoing threat posed by tanker ships plying the second busiest petroleum corridor in North America.

And, of course, the great, looming danger posed by global warming and rising sea levels, whose impact upon this low-lying region will be dramatic.

Like it or not, ready or not, the bayshore is on the verge of rapid evolution, and the only thing standing between the region's rich past and its uncertain future is the divided present.

"What would you like to see here, George?" I asked, indicating the pilings, the remains of the town. It was an academic question, made real by the person who answered it.

"A center or an organization, maybe," he said, thinking, at first, narrowly. "Something dedicated to keeping the bayman's heritage alive.

"Or a biological protection zone," he added, thinking more broadly, maybe grandly. "Something that will protect the bay, help us preserve the wisdom we have in order to protect the bay and the wildlife."

I nodded. Appreciating both the scope of the vision and the sentiment.

There are, and have been for several years, efforts to establish the Delaware Bayshore as a national heritage area—a federal designation that grants recognition to the natural and cultural value of a site or region. A "reconnaissance study" completed by the National Park Service in 2001 concluded that New Jersey's Delaware Bayshore met the standards for national heritage designation. There is every reason to believe that the bayshore will soon join the forty or so battlefields, historic towns, and culturally significant places that now enjoy national heritage status.

But the guiding limitations accorded such areas fall short of the oversight exercised in the adjacent Pinelands National Reserve—a region whose cultural heritage is likewise great and whose biological riches were also too important to lose. In truth and fact, the pine barrens and the bayshore are culturally and biologically joined, the bordering wetlands, the beaches, and the bay analogous to a superrich icing on a multilayered ecological cake. When the pine barrens was accorded protection, the icing was left off the cake.

If George had posed the question to me, instead of me to him, my response would have been similar. I, too, would like to see the bayshore's heritage—biological and cultural—protected.

But I would like to see future development restricted to existing villages—restoring their cultural significance and economic vitality, protecting surrounding natural areas from the sprawl that engulfed so much of North Jersey and now threatens South.

I would like to see these bayshore towns to become portals for visitors who, brought to awareness of the bay's exciting riches, will see it and become staunch advocates of the region, its people, and its environment.

I would like to see strict standards applied to outside lighting, so that the night sky will always be part of the dowry of residents and a treat for visitors.

I would like to see a biological master plan for the bay—one that makes Delaware Bay watermen and their four-hundred-year tradition an integral element to be incorporated, not just an interest to be accommodated.

I would like to see people like George Kumor and Tom Pew and Bill Garrison, and David Shepherd and maybe a few deer poachers, too (so that the Greg Honachefskys of the world can carry on their honorable profession), still here, still part of a living tradition when, one hundred years from now, another writer comes along who commits himself or herself to writing about the riches of the bayshore.

"Who's going to make it happen?" I asked George. "Give the bay the protection it deserves?"

"I'm wondering why 'you people' haven't stepped up to the plate," he said with typical Captain Kumor directness.

I guess I wonder, too. Why the future of New Jersey's last great treasure should still be uncertain when the price of doing nothing is just half a state away. Why so many good and well-intentioned agencies and organizations could be so focused on the bay and its environment yet overlook the biological element that makes the region so colorfully unique.

The people. Who have, for eight and ten and twelve generations, worked their changes upon and been themselves changed by the environment in which they live.

Not as observers but as participants. Not as exploiters but as players in a wonderful, real-life drama, involving people and nature. All set on an extraordinary stage that might be approaching its final curtain call, because the cast is dwindling, the directors are reading from conflicting scripts, and the audience seems not to know or care.

The tide was already turned when we started back in. On the water, you can sometimes gamble on the weather, but you can't play loose with the tide if you don't want you, your boat, and your catch to be stranded.

We reached the fixed end of the gill net we'd dropped earlier, and George began to haul, pulling hand over hand, piling the net in its waiting barrel, relieving it of the pearl-colored bunkers as they came over the side. This practice has changed little since men first set out to draw a harvest from the waters. Were Peter, or James or John, or any of Christ's apostles ever to cast nets in Delaware Bay, or were George transported back to the Sea of Galilee, they'd recognize each other as fellow tradesmen and appreciate one another's skill.

Or maybe there's a secret hand signal.

The haul produced few bunkers, fewer than thirty. These—two butterfish, a single small bluefish, and less than half a bushel of crabs—constituted the catch for the captain this day.

We'd been on the water five hours. It wasn't much to show. It certainly didn't support the contention George made as our boat nosed its way back into Kumor's Ditch.

"I'm optimistic about the future," he said. "We faced worse than this when the prison was discharging into the bay, and the bay came back."

It did. And it survived the pollution dumped into its waters by upstream industry during World War II. And the overharvest and diminishment of key marine animals over several hundred years. Nature is resilient; always compensating, always evolving, always trying to keep up with the changes foisted upon it. Human cultures are different. When cultures disappear, they are gone forever.

George was silent for a time, and when he did speak, he had to raise his voice over the sound of the engine, the tool that linked him to his trade, his culture, his bay. A workingman, a bayman, who not "from," but with "rude nature [had] delved and carved all the grandeur we behold."

"I'm just disappointed that people don't appreciate us more," he said, and I responded to this with silence. Not from smugness but to honor a truth, truthfully stated.

We were within sight of the landing now, and, surprisingly, there were people. A couple of kayakers putting in. A family led astray by a rusting highway sign but getting their bearings. And a guy standing right at the water's edge, who hailed us when we were within range of his voice.

"Hey, George," he shouted. "I need some information from the man who knows everything about the bay."

"Did you arrange this?" I chided, knowing it wasn't so.

George, hardly able to disguise his amusement, said, "No."

While George and the guy who wanted information about the ditches and tidal flow conversed, I started unloading the boat.

Wondering whether there was still enough time, will, and wisdom to wrap the whole Delaware Bayshore in a pine barrens reserve–like package to keep its heritage and biological integrity alive.

Wondering how long it's going to take for weakfish to come back and whether anyone with a bayman's skill will still be on the water to meet them.

Wondering where Harold, the night-heron, had gotten off to and whether he'd be back as soon as the crowd thinned.

Whether maybe this guy wasn't, after all, the E. F. Hutton of the Bayshore.

Anybody listening?

Note to Readers

No one who reads this book should conclude that mine is a lone voice when it comes to promoting the natural and cultural integrity of Delaware Bay. There are many fine and dedicated individuals associated with a host of organizations and institutions that are, in part or in full, committed to this ambition.

It is my privilege to commend the following groups on their efforts and to you: American Littoral Society, www.littoralsociety.org; Bayshore Discovery Project, www.ajmeerwald.org; Citizens United to Protect the Maurice River and Its Tributaries, Inc., www.cumauriceriver.org; National Park Service, New Jersey Coastal Heritage Trail Route, www.nps.gov/neje; Natural Lands Trust, www.natlands.org; the Nature Conservancy, www.nature.org; New Jersey Audubon Society, www.njaudubon.org; New Jersey Conservation Foundation, www.njconservation.org; New Jersey Division of Fish and Wildlife, www.state.nj.us/dep/fgw; and, of course, the governing councils of the region's towns and municipalities who have, since their incorporation, worked to promote the interests of the Bayshore and its residents.

Bibliography

Alexander, Robert Crozer. *Ho! For Cape Island!* N.p.:
    Edward Stern & Company, 1956.

Britt, Robert Roy. "Meteors and Meteor Showers: The
    Science." http://www.space.com/scienceastronomy/
    solar system/meteors-e2.html.

Carman, Alan E. *The Lenape Culture*. Millville: Cumber-
    land County Prehistorical Museum, n.d.

*Commercial Township Salt Hay Farm Wetland Restoration
    Management Plan*. N.p.: PSE&G, 1995.

Cox, Sam. "I Say Tomayto, You Say Tomahto. . . ."
    http://www.landscapeimagery.com/tomato.

Fox, William T. *At the Sea's Edge*. New York: Prentice Hall
    Press, 1983.

"History of Agriculture in Cumberland County." http://
    www.co.cumberland.nj.us/content/161/223/373
    /default.aspx.

"In Search of the *Real* Jersey Tomato," Part 1. http://njfarm
    fresh.rutgers.edu/pdfs/whatsinseason7-31.pdf.

"In Search of the *Real* Jersey Tomato," Part 2. http://njfarm
    fresh.rutgers.edu/pdfs/whats-in-Season-8-14.pdf.

Knowlton, Clarence Hinkley. "Dallas Lore Sharp—An
      Appreciation," *Bostonia* 13, no. 9, n.d., pp. 3–9.
"Light Pollution." http://en.wikipedia.org/wiki/Light
      _pollution.
The Lower Maurice River Resource Identification Project.
      Media, Pa.: Natural Lands Trust, 1993.
*The Maurice River and Its Tributaries National Wild
      and Scenic River Study Draft Report.* Philadelphia:
      National Park Service, 1992.
"Meteor Shower Basics." http://www.amsmeteors.org/
      showers.html#basics.
"Meteoroid." http://en.wikipedia.org/wiki/Meteor#
      Meteoroid.
Moskin, Julia. "The Return of a Lost Jersey Tomato."
      http://www.nytimes.com/2008/07/23/dining/23toma
      .html.
Reddy, Francis. "Meteors and Meteor Showers." http://
      astronomy.com/asy/default.aspx?c=a&id=2109.
Reeves, Joseph S. *Maurice River Memories– Cumberland
      County, New Jersey 1937–1947.* Baltimore: Gateway
      Press, 1993.
Rey, H. A. *The Stars—A New Way to See Them.* Boston:
      Houghton Mifflin Company, 1952.
Rosshart, Verna. "Dallas Lore Sharp." *Bridgeton Evening
      News*, n.d.
Sebold, Kimberly. *From Marsh to Farm: The Landscape
      Transformation of Coastal New Jersey.* Washington,
      D.C.: National Park Service, 1992.
Sebold, Kimberly, and Sara Amy Leach. *Historic Themes*

and Resources within the New Jersey Coastal Heritage Trail. Washington, D.C.: U.S. Department of the Interior, National Park Service, 1991.

Sharp, Dallas Lore. *Summer*. Cambridge, Mass.: Riverside Press, 1913.

Stone, Wes. "Major Meteor Showers in 2008." http://skytour.homestead.com/met2008.html.

Strahler, Arthur N. *Physical Geography*. New York: John Wiley and Sons, 1951.

"Tomato." http://en.wikipedia.org/wiki/Tomato.

Williams, Karen. *"Bitey" Things*. Bernardsville: New Jersey Audubon, 1990.

Wood, Jonathan E. *The Cumberland Story—A Brief History of Cumberland County, N.J.* Millville: Cumberland County Historical Society, 1986.

Worth, C. Brooke. *Of Mosquitoes, Moths, and Mice*. New York: W. W. Norton & Company, 1972.